THE PILOT'S OUTLINE GUIDE TO BASIC AERODYNAMICS

William M. Jones

First Edition
1995
Eastern Dakota Publishers

Edited by Ronald G. Gallagher
Graphics by Steve Pottenger

ISBN 0-9639984-2-0

Published by:

Eastern Dakota Publishers
P.O. Box 12512
Grand Forks, ND 58208-2512

First Printing

Printed by:

Century Creations Printing Inc.
1440 South Washington
Grand Forks, ND 58201

Warning/ Disclaimer

Any procedure or practice described in this book should be applied by a properly rated pilot, under appropriate supervision in accordance with professional standards and regulations with regard to the unique circumstances that apply in each situation. Care has been taken to confirm the accuracy of information presented and to describe generally accepted practices. However, the author, editor, and publisher cannot accept any responsibility for errors or omissions or for the consequences from application of the information in this book and make no warranty, express or implied, with respect to the contents of the book.

About the Author

William M. Jones has been flying aircraft since 1973. He is a retired Marine Corps LtCol. and served as a Naval Flight Officer with a specialty as a Bombardier/ Navigator in the A6 Intruder. After retiring from the Marine Corps he was hired as an Assistant Aviation Education Specialist and Flight Instructor for the University of Illinois, where he currently teaches. William is also the Senior Flight Instructor for the Illini Glider Club. He holds an Airline Transport Pilot, Commercial Pilot, Certified Flight Instructor- Single engine, Instrument, Multi engine, and Glider, along with Advanced and Instrument Ground Instructor ratings.

The author wishes to extend particular gratitude to Rick Weinberg and Steve Owen from Pilot Training at the University of Illinois for their editorial comments on the original manuscript. A special thanks goes to Don Talleur from the Aviation Research Lab and Pilot Training at the University of Illinois for his editorial comments and his continued professional encouragement in completing this book.

Preface

The purpose of this book is to focus entirely on aerodynamics which would be user friendly to all new aviation students and also serve as an instructional guide for flight instructors. It has been my observation that beginning flight students may not have a solid grasp of basic aerodynamics, thus creating possible problems when these pilots are working towards advanced pilot ratings like the commercial or flight instructor certificates. I feel the problem is due in part because aerodynamic principles are either contained in an overly complicated textbook (designed primarily for aeronautical engineers) or are only given partial treatment in general aviation books. This book, therefore, has been written to fill the gap between the overly simple and the overly complex, and will hopefully provide a solid foundation of the basics.

TABLE OF CONTENTS

Chapter 1

Aircraft Nomenclature and the Atmosphere

Aircraft Motion:

✈ All motions of an aircraft take place about the center of gravity (CG). It is the point at which a body would balance if placed on a fulcrum. (Reference Fig. 1-1)

✈ Types of motions about the center of gravity:
1. Yaw: side to side movement of the nose about the vertical axis.
2. Pitch: up and down movement of the nose about the lateral axis.
3. Roll: the movement of the wings about the longitudinal axis.
(Reference Fig. 1-2)

Aircraft Components:

✈ Three major components on an aircraft:
1. Fuselage (main body).
2. Wings (provide lift).
3. Empennage (tail assembly).

✈ Ailerons:
1. Pair of movable control surfaces located on the outboard trailing edge of the wing.
2. Move in opposite direction to each other.
3. Controls roll about the longitudinal axis.

✈ Vertical and horizontal stabilizers: fixed surfaces making up a major part of the empennage.

✈ Rudder:
1. A movable control surface at the trailing edge of the vertical stabilizer.
2. Controls yaw about the vertical axis.

✈ Elevators:
1. Movable control surfaces located at the trailing edge of the horizontal stabilizer.
2. Controls pitch about the lateral axis.

✈ Trim Tabs: attached to the trailing edge of control surfaces. They aid in the control of the aircraft by reducing the control forces required of the pilot.

Airfoil Terminology:

(Reference Fig. 1-3)

✈ Chord: the average straight line distance between the leading and trailing edges of an airfoil (c).

✈ Chordline: the straight line which forms the chord. Can be extended ahead of and behind the airfoil.

✈ Mean Camber Line: The line formed by connecting all points which are equidistant from the upper and lower surfaces of an airfoil.

1. Symmetrical airfoil: mean camber line and the chordline are the same. Horizontal and vertical stabilizers are examples.
2. Positive camber: mean camber lies above the chordline.
3. Negative camber: mean camber lies below the chordline.

✈ Flight Path (FP): the line which traces the direction of travel of the center of gravity through the air mass.

✈ Relative Wind (RW): the wind created by the direction of travel and the speed of the aircraft through the air mass. It is equal in velocity and opposite in direction to the flight path velocity (TAS).

✈ Angle of Attack (∝): Angle between the relative wind or flight path and the chordline. Greek letter Alpha (∝).

✈ Angle of Incidence: the angle formed by the chordline of the wing and the longitudinal axis of the aircraft. It is normally a fixed angle.

✈Pitch Angle: the angle formed by the horizon and the longitudinal axis.

✈ Wing Span (b): the physical length of the wing.

✈ Wing Area (S): the product of the wing's span (b) and the wing's average chord (c). Includes the part of the wing that goes through the fuselage.

✈ Aspect Ratio (AR): the ratio of the wing span to it's average chord.

$$AR=b/c$$

✈ Forces Working on an Aircraft: lift, weight, drag, thrust. (Reference Fig. 1-4)

Newton's Laws:

✈ First Law: law of inertia or the law of equilibrium.

1. A body at rest tends to remain at rest and a body in motion tends to remain in motion until acted upon by an unbalanced force.
2. Therefore, the forces on a non-maneuvering aircraft must sum up to zero. (thrust=drag; lift=weight)
3. All the moments about an aircraft in equilibrium will be zero if the aircraft is not under going some type of acceleration.
4. Therefore, an aircraft at a constant speed and in a constant direction of flight will be in equilibrium. This also applies to aircraft climbing or diving at a constant airspeed.

✈ Second Law: law of acceleration.

1. F=MA. When there exists an unbalance of forces , some resultant force acts on the aircraft and an

acceleration (speed up, slow down, turn, etc.) is produced and the aircraft is no longer in equilibrium.

✈ Third Law: law of motion.
1. For every action there is an equal and opposite reaction.

The Atmosphere:
✈ Static Pressure (Ps):
1. Weight of a column of air over a certain area.
2. As altitude increases, static pressure decreases.

✈ Temperature (T):
1. Measure of the average molecular energy of the air.
2. Temperature decreases as altitude is increased until approximately 36,000 feet where it remains constant at -59° C or -70° F.

✈ Density:
1. Mass per unit volume. Greek symbol rho (ρ).
2. As altitude increases, density decreases. Example: at 40,000 feet the atmospheric density is only 1/4th as great as it is at sea level.

✈ General Gas Law:
1. States that density of a gas varies directly with pressure and inversely with temperature.
2. Increases in humidity (amount of water vapor in the air) also decrease density.

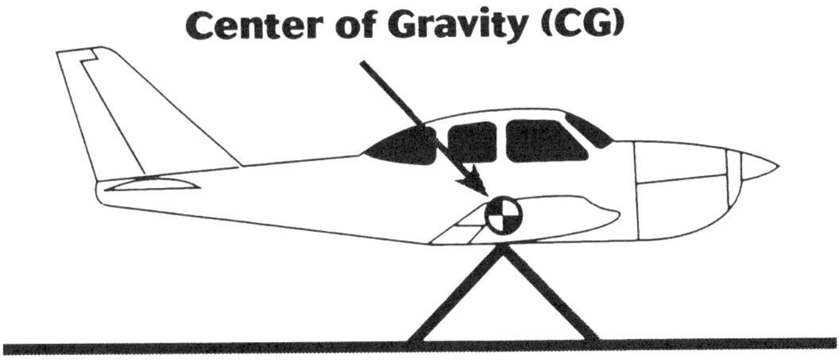

Figure 1-1

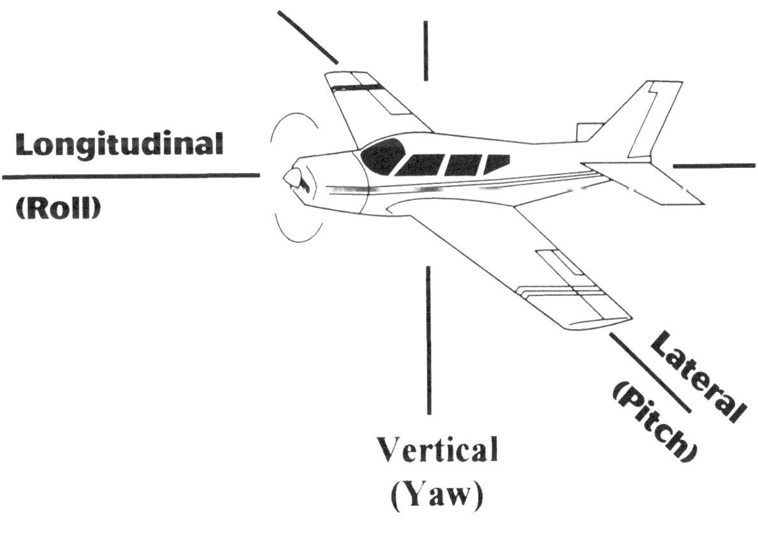

Figure 1-2

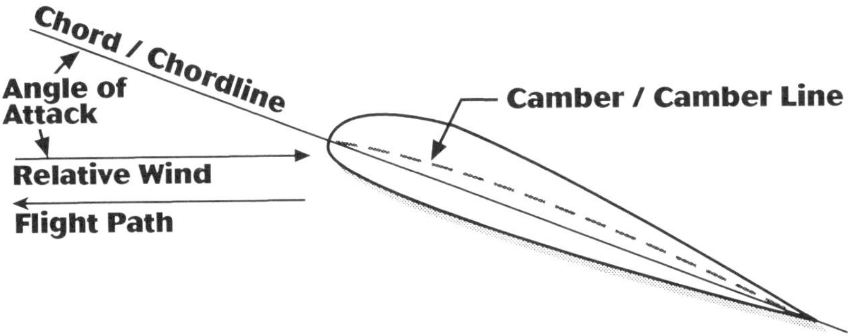

Figure 1-3

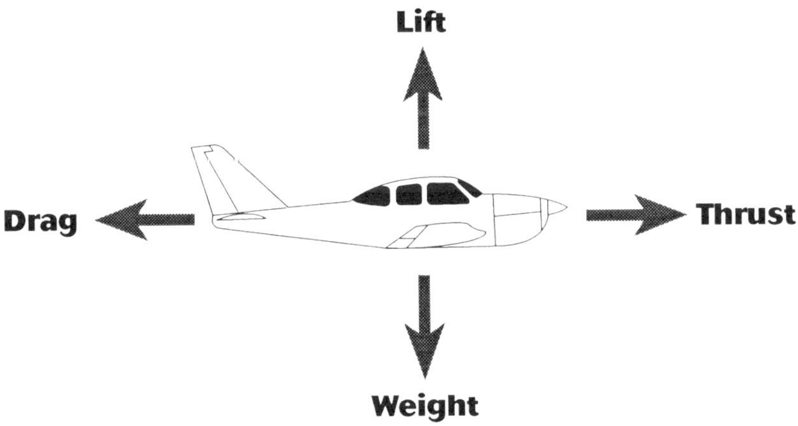

Figure 1-4

✈ Density Ratio:
1. Greek letter sigma (σ).
2. Used because the unit for density which is slugs per cubic foot is difficult to work with.
3. Is the ratio of the density at any altitude divided by the density at sea level ($\rho 0$).

$$\sigma = \rho / \rho 0$$

Study Questions Chapter 1

1. All motions of an aircraft take place about what?

2. Describe yaw.

3. Describe pitch.

4. Describe roll.

5. What are the three major components of an aircraft?

6. What control surface is attached to the trailing edge of the wing?

7. Ailerons control what movement about which axis?

8. What two major sections make up the empennage?

9. What control surface is located at the trailing edge of the vertical stabilizer?

10. The rudder controls what movement about which axis?

11. Where is the elevator located?

12. The elevator controls what movement about which axis?

13. What do trim tabs do?

14. Describe chord.

15. Describe chordline.

16. Describe mean camber line.

17. Describe flight path.

18. Describe relative wind.

19. Describe angle of attack.

20. Describe angle of incidence.

21. Describe pitch angle.

22. Describe positive camber.

23. Describe negative camber.

24. What is a symmetrical airfoil?

25. The horizontal and vertical stabilizers are most often what type of an airfoil?

26. Describe wing span.

27. Describe wing area.

28. Describe aspect ratio.

29. Describe Newton's first law of equilibrium.

30. What will all the moments about an equilibrium aircraft be if the aircraft is not under going some type of acceleration?

31. Describe Newton's second law.

32. What happens when an unbalance of forces exists on an aircraft?

33. Describe Newton's second law.

34. Describe static pressure.

35. What happens to pressure with an increase in altitude?

36. Describe temperature.

37. What happens to temperature with an increase in altitude?

38. At what altitude does temperature stabilize?

39. Describe density.

40. What happens to density with an increase in altitude?

41. How are pressure and density related?

42. How are temperature and density related?

43. How are humidity and density related?

44. What is the general gas law?

45. What is density ratio?

NOTES:

Chapter 2

Fluid Flow and Airspeed Measurement

Fluid Flow:

✈ Air is described as a fluid and therefore is subject to the laws of fluid motion.

✈ Continuity Equation for a closed tube, the mass flow of air into the tube must be equal to the mass flow out of the tube.

1. (pAV)in = (pAV)out

A= cross sectional area

V= velocity

p= mass

✈ If we say that no mass can cross the boundary of a tube anyplace except at the entrance or exit, and neglect changes in density, then:

1. (AV)in = (AV)out

✈ Since (AV)in = (AV)out, a decrease in area causes a velocity increase and an increase in area causes a velocity decrease.

(Reference Fig. 2-1)

✈ (Reference Fig. 2-2)
1. Va and Vb represent the boundaries of a tube and the airfoil represents a decrease in area. Therefore, when V1 encounters the leading edge of the airfoil, it senses a decrease in area and velocity will increase as it passes over the upper and lower surfaces of the airfoil.
2. Velocity will decrease as it moves towards the trailing edge because of an increase in area.
3. Also at the point where V1 strikes the leading edge of the airfoil, velocity will decrease to zero.

✈ Law of Conservation of energy states that for any closed system the total energy is a constant.

✈ Total energy is the sum of kinetic and potential energy.

Total Energy = Kinetic Energy + Potential Energy

(Kinetic means in motion, Potential means at rest)

✈ The amount of kinetic or potential energy can change, but the total remains constant.

✈ Equation becomes:

Total Pressure = Kinetic Pressure + Potential Pressure

✈ Static Pressure (Ps) was defined as the weight of a column of air, therefore, static pressure is the measure of potential pressure.

→ From physics, kinetic energy is written:

$$KE = 1/2MV^2$$

→ By substituting the atmospheric expression for density (or mass), sigma, and a conversion factor, dynamic pressure becomes:

$$\text{Dynamic Pressure (q)} = \frac{\sigma V^2}{295}$$

→ Bernoulli's equation of total pressure:

$$\text{Total Pressure } (H_t) = q + Ps$$

or

$$H_t = \frac{\sigma V^2}{295} + Ps$$

→ (Reference Fig. 2-3)

1. Air speeds up above and below the airfoil.
2. Increase in speed increases dynamic pressure and reduces static pressure.
3. Since the air speeds up more over the top of an airfoil than the bottom, static pressure decreases more on the top.

(Reference Fig. 2-4)

4. The pressure differential between the lower pressure on the top and the higher pressure on the bottom of the airfoil causes approximately 75% of the generated lift, with the remainder of lift being generated by the air that is deflected downward due to the airfoil's shape (Newton's Third Law of Motion).
5. As the air approaches the trailing edge of the airfoil, it slows down and the static pressure again rises.

6. Increases in static pressure at the leading and trailing edges of an airfoil are responsible for producing drag.

Airspeed Measurement:

✈ General:

1. Necessary to measure airspeed to determine when there is enough dynamic pressure to generate enough lift to sustain flight.
2. Necessary to measure airspeed for navigational purposes.
3. Since rapid changes in static pressure are difficult to measure, measuring dynamic pressure which varies with velocity over the wing gives the pilot a good indication of the aircraft's ability to produce lift.
4. By re-arranging Bernoulli's equation for total pressure, dynamic pressure can be found by:

$$q = H_t - Ps$$

✈ Pitot-Static System:

1. Aircraft use a pitot-static system to measure dynamic pressure and it is read in the cockpit in knots of airspeed, normally indicated airspeed (IAS). Older aircraft have used miles per hour.

Pitot-static system:

a. Pitot tube: placed in the airstream so that its internal back is perpendicular to the free air stream. As the airstream hits the back of the tube the velocity decreases to zero. This converts all the kinetic energy of the airstream to potential energy, therefore, the pressure measured by the pitot tube is total pressure.

b. Static port: located on the aircraft so the free airstream past it is parallel to the opening and thus is normally unaffected by the motion of the air, therefore, measures only the static pressure for the airstream. Changes in the static pressure occur by changing altitude.
c. Total system: Air pressure from the pitot tube and static port are fed into opposite sides of a chamber which is connected by a flexible diaphragm. The diaphragm measures changes in dynamic pressure and through a system of gears and levers transmits the movement of the diaphragm to an indicator in the cockpit. (Reference Fig. 2-5)

✈ Indicated Airspeed (IAS):
1. IAS is the measurement of dynamic pressure indicated in the cockpit on the airspeed indicator.
2. IAS is not an airspeed, simply a measurement of dynamic pressure.

✈ True Airspeed (TAS or V):
1. The actual velocity of an aircraft through the air mass in which it is operating.
2. Is equal to the velocity of the relative wind.
3. TAS is determined from IAS by the density ratio.
4. TAS only equals IAS when the density ratio is one (sea level).
5. As you increase altitude, with constant IAS, the TAS increases because the density ratio decreases.
6. TAS is not effected by wind.

✈ Ground Speed (GS):

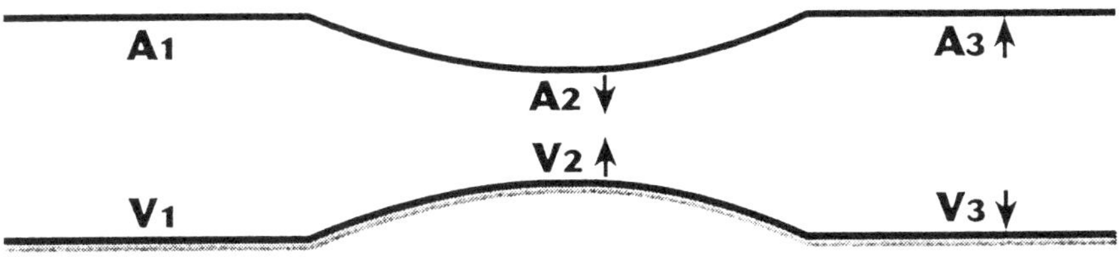

Figure 2-1

Figure 2-2

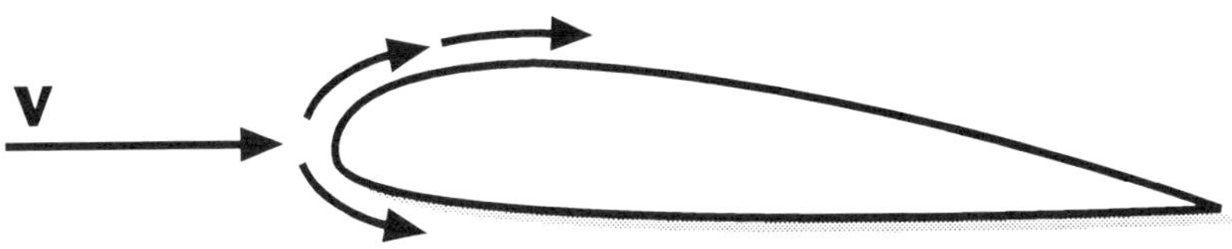

Figure 2-3

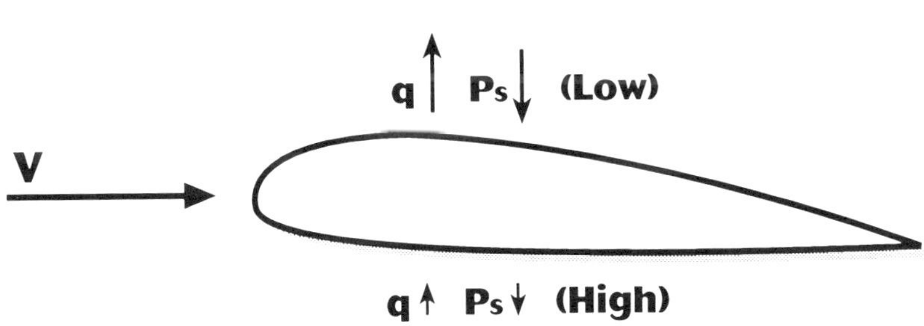

Figure 2-4

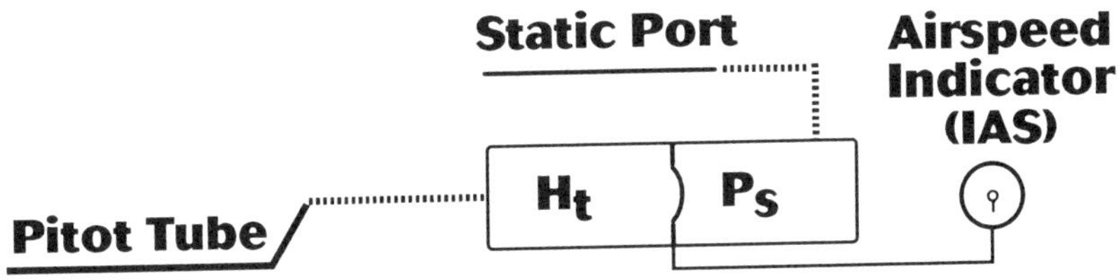

Figure 2-5

1. Speed of an aircraft relative to the ground.
2. Is effected by wind.

→ Mach Number:
1. Ratio of an aircraft's TAS to the speed of sound (Va) at your altitude.
2. Speed of sound is directly proportional to temperature. Since temperature generally decreases with an increase in altitude, the speed of sound decreases with an increase in altitude.

Study Questions Chapter 2

1. State the Continuity Equation.

2. Considering the Continuity Equation, when an area decreases, what happens to velocity?

3. Considering the Continuity Equation, when an area increases, what happens to velocity?

4. What happens to the total energy within a closed system?

5. What two elements make up total energy?

6. Total energy equals what?

7. Total pressure equals what?

8. What is the formula for dynamic pressure?

9. Write the total pressure equation two ways.

10. What happens to the velocity of air as it passes above and below an airfoil?

11. What happens to the total pressure equation above and below an airfoil?

12. Why is the static pressure less on the top of an airfoil than on the bottom?

13. What is responsible for creating lift?

14. What happens to velocity and static pressure at the leading and trailing edges of an airfoil?

15. What does the the increase in static pressure at the leading and trailing edges of an airfoil produce?

16. What are the two reasons for measuring airspeed?

18. Re-arrange Bernoulli's equation to express dynamic pressure?

19. What do pitot-static systems measure?

20. The pitot-static system causes the cockpit airspeed indicator to read out in knots, but it is actually measuring what?

21. The measurement of dynamic pressure that the pilot reads in the cockpit is called what?

22. Is IAS actually an airspeed?

23. What is the actual velocity of the aircraft through the air mass in which it is operating?

24. What is true airspeed exactly equal to?

25. What is needed to convert IAS to TAS?

26. Is TAS effected by wind?

27. What happens to TAS when holding a constant IAS in a climb or descent?

28. Describe Ground Speed.

29. Is ground speed effected by wind?

30. Define Mach number.

NOTES:

NOTES:

Chapter 3

Lift, Stalls, and Turning Flight

Lift:

✈ Aerodynamic Force:
1. Total force acting on an airfoil.
(Reference Fig. 3-1)
2. Difficult to measure because it varies greatly with changes in angle of attack and velocity.
3. Therefore, aerodynamic force is broken down into two component forces.
 a. Lift: perpendicular to the RW or FP.
 b. Drag: parallel to RW or FP.

✈ Lift Equation: $L = qSC_L$ or $L = \frac{\sigma V^2}{295} SC_L$

1. C_L = Coefficient of lift (coefficient is a number expressing the amount of some change and C_L is the ratio of lift pressure and dynamic pressure).
2. C_L is a function of angle of attack since the camber of the wing is fixed.
3. C_L increases linearly with angle of attack until a certain value is reached and then drops off rapidly as angle of attack is further increased.
(Reference Fig. 3-2)

4. C_L MAX: Coefficient of lift maximum is the largest value for C_L before the aircraft stalls.
   a. Is called the stalling angle of attack (αs).
   b. Determines minimum airspeed.
5. Lift developed by a wing: only quantities in the lift equation which effect lift are:
   a. q = dynamic pressure.
   b. C_L = coefficent of lift.
6. Therefore, lift can be increased or decreased by changing q or C_L.
7. Changes in q are caused by changing TAS or density.
8. Changes in C_L are created by changes in angle of attack.
9. To maintain level flight, an aircraft must create lift equal to its weight.
10. To maintain level flight at a higher speed, C_L (angle of attack) must be decreased as the airspeed increases.

$$L = q(\uparrow)\ S(\text{constant})\ C_L\ (\downarrow)$$

11. To maintain level flight at a slower speed, angle of attack must be increased.

$$L = q(\downarrow)\ S(\text{constant})\ C_L\ (\uparrow)$$

Stall:

✈ General:

1. Stall is defined as a loss of lift.
2. Occurs to the right of C_L MAX where a decrease in C_L occurs with an increase in angle of attack. (Reference Fig. 3-3)
3. *Cause of stall is excessive angle of attack.*
4. *To recover from a stall, must reduce the angle of attack.*

✈ What happens to cause a stall:

1. At zero or low angles of attack, subsonic air flows smoothly around an airfoil.
(Reference Fig. 3-4)

2. With increasing angles of attack, the smooth air flow begins to break away from the curvature of the airfoil near the trailing edge.
(Reference Fig. 3-5)

3. The turbulent air doesn't change pressure with changes in velocity like smooth air, so it doesn't produce lift.

4. As the angle of attack is increased, the turbulent air moves forward toward the leading edge of the wing.

5. At the stalling angle of attack, enough of the wing is effected by the turbulent air that the wing cannot generate enough lift to balance weight and the aircraft stalls.

✈ Control of Stalls:

1. Stalls should occur at the wing root first instead of the whole wing at once.

   a. This is so the turbulent air created before the stalling angle of attack is reached will hit the horizontal stabilizer first and give the pilot a stall warning.

   b. Ailerons are near the wing tips, so by having the wing root stall first, the ailerons are still effective for lateral control during the stall recovery if recovery is made at the first indication of a stall.

✈ Methods for causing stalls to occur at the wing root:
1. Geometric Twist: A twisted wing where the wing's angle of incidence is greater at the wing root than at the outer portion of the wing towards the wing tip.
2. Aerodynamic Twist: Wing is not twisted but is built of two separate airfoil shapes. A symmetrical airfoil stalls at a higher angle of attack than a cambered airfoil. Therefore, the wing root will be cambered and the wing tip portion will be symmetrical.
3. Stall Strip: Small strip attached to the leading edge of the wing root, and causes boundary layer separation at a lower angle of attack.
4. Stall Fence: Used on swept wing aircraft and is a metal strip placed parallel to the free air stream on the top of the wing. Swept wing aircraft have a tendency to stall at the wing tips due to "spanwise flow", with spanwise flow noticeable at high angles of attack. Stall fence reduces the spanwise flow.

✈ Stall Warning Devices:
1. Flapper Switch: Flapper switch is mounted on the leading edge of a wing. Is used mainly on civilian aircraft and is activated either electrically or by air pressure.
2. Angle of Attack System: Used mainly on larger aircraft such as military jets and consists of a probe positioned on the fuselage. It aligns itself with the relative wind with the indicator in the cockpit calibrated in units rather than degrees of angle of attack. The system usually consists of a rudder or stick shaker system to warn the pilot.

Turning Flight:

✈ General:

1. To maintain a level turn, lift must equal weight.
2. When an aircraft goes into a turn, the lift vector is divided into two components:
   a. Vertical lift component (Lv).
   b. Horizontal lift component (Lh).

(Reference Fig. 3-6)

3. Vertical component of lift must equal weight to maintain a level turn.
4. At any angle of bank where the vertical component of lift is insufficient to balance weight the total lift vector (Lt) must be increased so that the vertical component can be increased to balance weight or else the aircraft will not be able to maintain altitude during a turn.
5. Looking at the lift equation, total lift can be increased by increasing speed (q) or C_L (angle of attack).
6. Considering a constant speed turn, the only way to increase lift is to increase C_L by increasing the angle of attack.
7. As the angle of bank increases for a turn, the total lift vector must be increased to maintaina level turn.
8. At a 60 degree angle of bank, twice as much total lift is required for the vertical component of lift to equal weight.

✈ Load Factor:

1. The ratio of the total load supported by the wing to the actual weight of the aircraft (N).

$$N = \frac{Lt}{W}$$

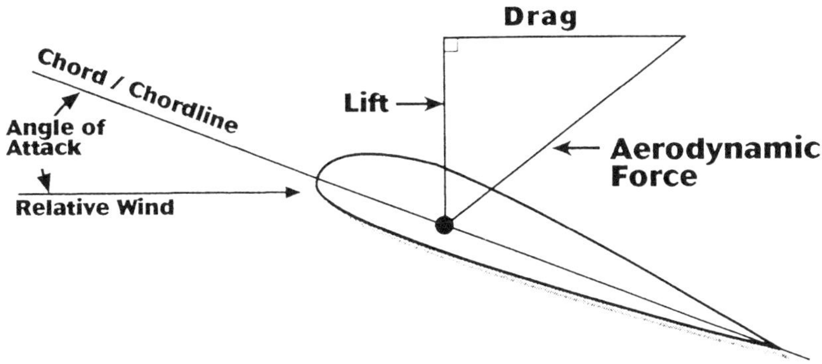

Figure 3-1

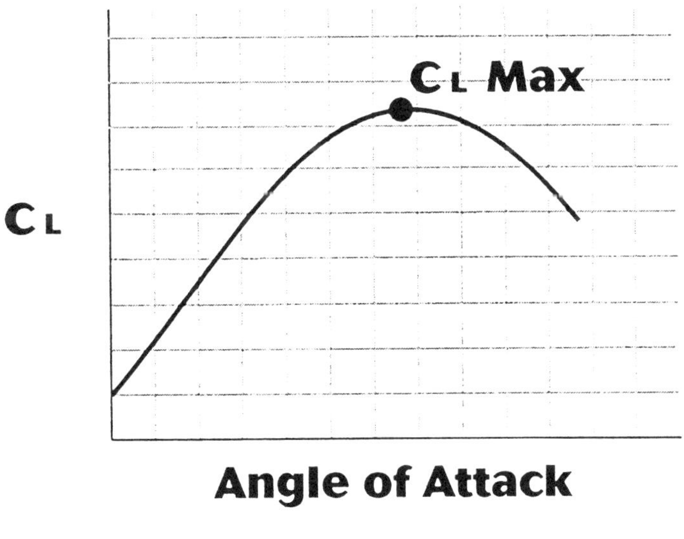

Figure 3-2

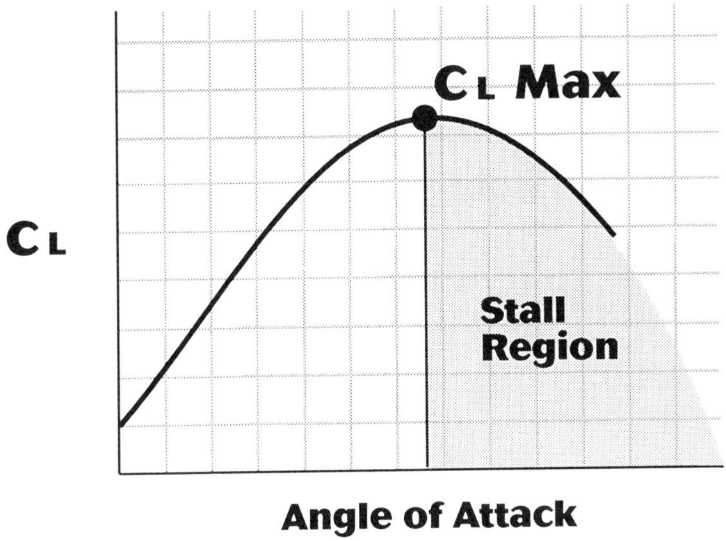

Figure 3-3

Figure 3-4

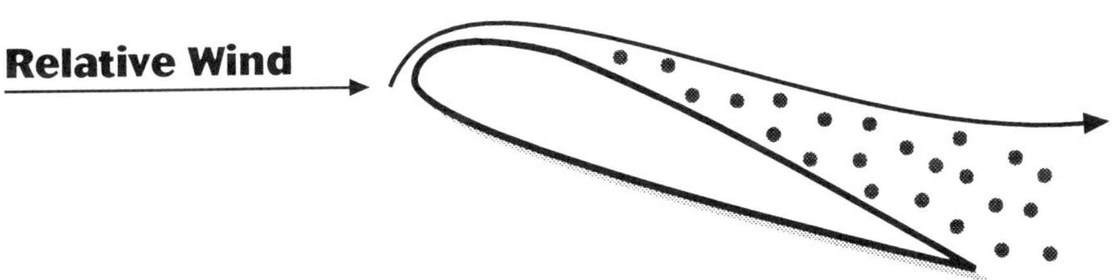

Figure 3-5

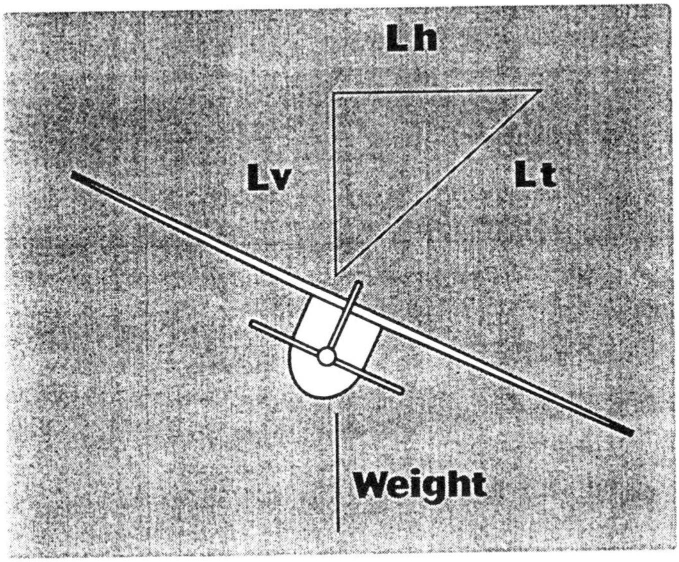

Figure 3-6

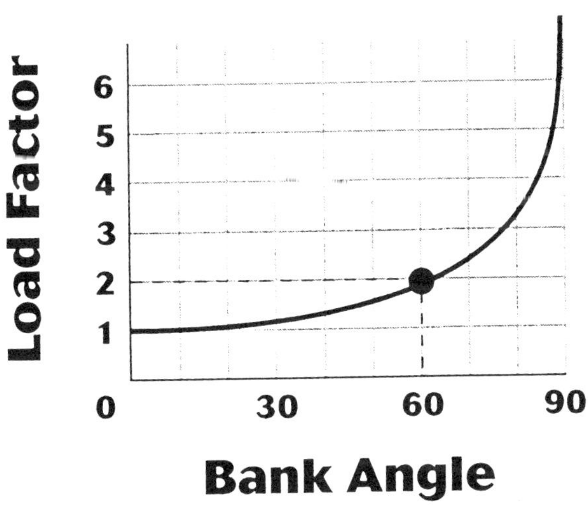

Figure 3-7

2. At a 60 degree angle of bank there is a load factor of 2.
(Reference Fig. 3-7)
3. A pilot feels the load factor by a change in his weight and is called "G's".
4. Radius of turn: The size of the turn, or radius of the turn, is flown at the slowest possible speed and at the highest angle of bank obtainable.
5. Rate of Turn: Number of degrees of the arc of a circle traversed by the nose of an aircraft in a given amount of time. Also, flown at the slowest possible speed and the highest possible angle of bank.
6. Level Turn Limitations:
    a. Power available: Power is required to overcome drag. Drag increases as angle of attack in a turn increases so power is needed to overcome the drag and maintain a constant speed turn.
    b. Stalling angle of attack.
    c. Structural strength of the aircraft.

Study Questions Chapter 3

1. What are the two component forces of the aerodynamic force?

2. Define lift.

3. What is the coefficient of lift a function of?

4. What is C_L MAX?

5. What are the only two quantities which can effect lift?

6. How can lift be increased?

7. What is needed to maintain level flight?

8. What is needed to maintain level flight at an increased airspeed?

9. What is needed to maintain level flight at a reduced airspeed?

10. What does a decrease in C_L represent?

11. Define a stall.

12. What causes a stall?

13. How do you recover from a stall?

14. Why do we try to get stalls to occur at the wing root first?

15. Describe geometric twist.

16. Describe aerodynamic twist.

17. Describe a stall strip.

18. What is a stall fence?

19. What is needed to maintain a level turn?

20. In a constant speed turn, what is the only way for the pilot to increase lift?

21. What is load factor?

22. What is an aircraft's radius of turn a function of?

23. How would a pilot fly the minimum radius turn?

24. Define rate of turn and how would you fly the greatest rate of turn?

25. Describe the three limitations to level turns.

NOTES:

NOTES:

Chapter 4

Drag

General:

✈ Drag is the retarding force on an aircraft that acts parallel to the relative wind.

✈ Drag is opposed by thrust.

✈ There are two types of drag. Their total equals total drag (Dt):

1. Parasite Drag (Dp).
2. Induced Drag (Di).

Parasite Drag:

✈ All drag not associated with lift. There are three major types of parasite drag:

1. Form: drag produced by the pressure build up on a body due to its shape.
2. Skin Friction: develops as air passes over a skin surface
3. Interference: forms where two or more component airflows come together, such as the junction of a wing and the fuselage.

✈ Equation for parasite drag:

Dp = qf (f = equivalent parasite area)

or

$Dp = (\sigma v^2 / 295)\ f$

✈ An increase in velocity creates an increase in parasite drag.
✈ Parasite drag varies directly with velocity squared. (Reference Fig. 4-1)

Induced Drag:
✈ Induced drag is all drag associated with the production of lift.
✈ Molecules on the bottom of the wing try to move to the low pressure area on the top of the wing. Molecules can't go through the wing or around the leading or trailing edges due to the high static pressure built up. Therefore, they can only reach the top of the wing at the wing tips. Air from under the wing moves upward around the wing tip and is caught in the flow of air moving around the wing from the leading to the trailing edge. When the two airflows mix, the result is a turbulent, circular airflow called wing tip vortices.
✈ Wing tip vortices are the actual cause of induced drag.
✈ Induced drag varies inversely with velocity squared. (Reference Fig. 4-2)
✈ Induced drag is the greatest at low speeds and at high angles of attack.
✈ Methods of reducing induced drag:

1. Increase a wing's aspect ratio (longer wing). This makes it more difficult for the air under the wing to reach the upper surface (sailplanes use this technique).
2. Place tip fuel tanks on the end of the wings.
3. Bend the tips of the wings up or down. (example = winglets)

Coefficient of Drag:

✈ $D = qSC_D$

✈ Coefficient of drag (C_D) is a function of angle of attack and camber.

✈ Since camber is fixed, C_D increases with an increase in angle of attack.

(Reference Fig. 4-3)

✈ C_D plotted against C_L.

(Reference Fig. 4-4)

Total Drag:

✈ Dt = Dp + Di

✈ Total drag can be represented by adding parasite and induced drag.

(Reference Fig. 4-5)

✈ Total drag starts out high at low airspeeds due to Di and drops to a minimum value where the sum of Dp and Di are minimum and then increases with velocity increases due to Dp.

✈ Point where Dt is minimum is important for several reasons:

1. Airspeed where drag on the aircraft is minimum.
2. Point where the aircraft is operating most efficiently (L/D – lift divided by drag. L/D is the measure of the amount of lift produced by a wing while producing a certain amount of drag).

L/D Max:

✈ L/D Max is the bottom point on the total drag curve. The wing produces a maximum amount of lift for a minimum amount of drag. *The wing is operating at its most efficient point at L/D Max.*

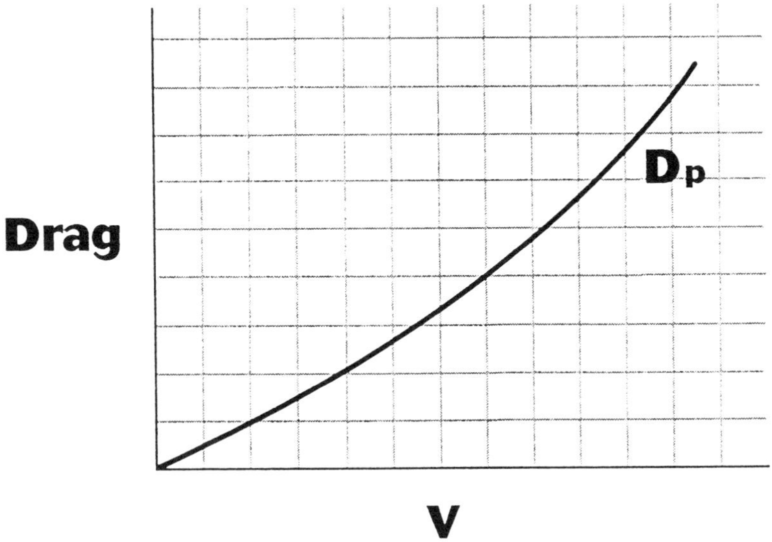

Figure 4-1

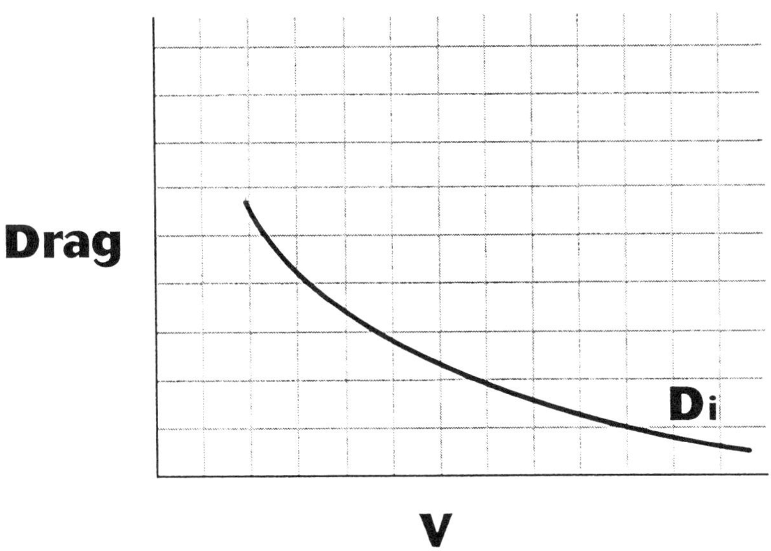

Figure 4-2

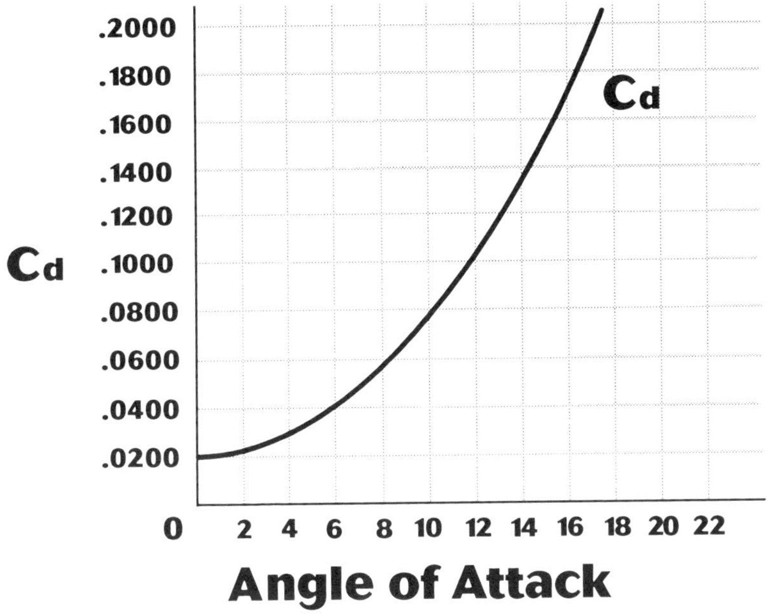

Figure 4-3

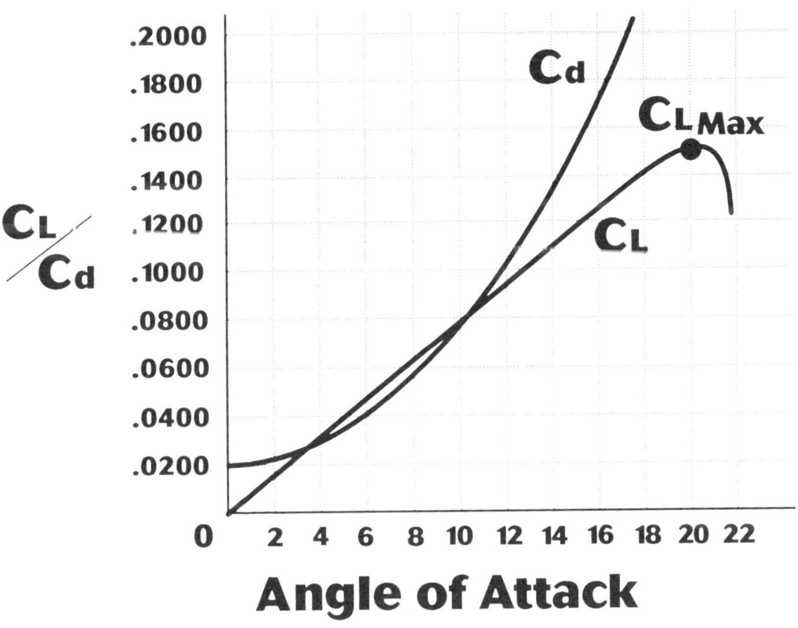

Figure 4-4

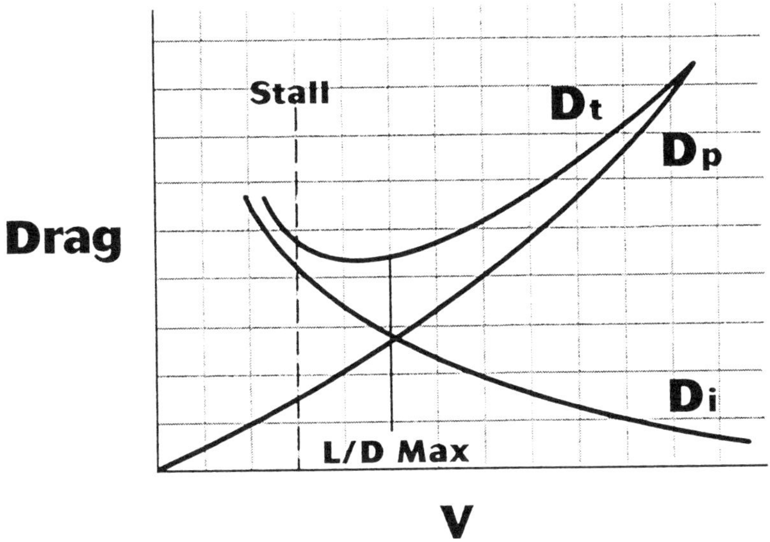

Figure 4-5

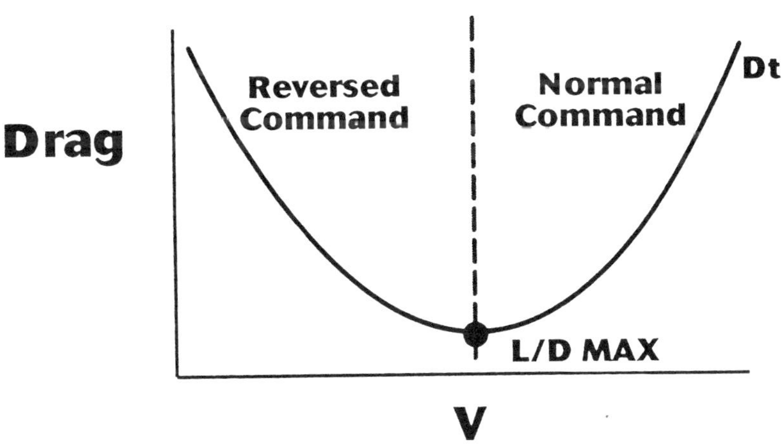

Figure 4-6

Region of Normal and Reversed Command:

✈ L/D Max divides Dt drag graph into two regions. (Reference Fig. 4-6)

✈ Region of normal command: to the right of L/D Max. To maintain straight and level flight, thrust must equal drag. To slow down, reduce thrust.

✈ Region of reversed command: to the left of L/D Max. Almost all of the drag is induced drag. To slow down and maintain level flight you must increase thrust to counter the increase in total drag. Trying to fly slower (back stick) without more thrust would cause a still slower airspeed caused by more total drag (mainly Di). This is the reason why a glide cannot be stretched by only pulling back on the yoke/ stick. The slower airspeed without an increase in required thrust to overcome the increased drag then causes the aircraft to pick up a higher rate of descent and therefore not glide as far.

Study Questions Chapter 4

1. Define drag.

2. What opposes drag?

3. What are the two types of drag?

4. Define parasite drag.

5. What are the three major categories of parasite drag?

6. Define form drag.

7. Define skin friction.

8. Define interference drag.

9. How are velocity and parasite drag related?

10. Define induced drag.

11. What is the actual cause of wingtip vortices which is the cause of induced drag?

12. How are velocity and induced drag related?

13. What are three methods for reducing induced drag?

14. Define and express total drag.

15. Draw the graph for total drag.

16. Why is the point on the total drag curve where drag is at a minimum important?

17. Define L/D.

18. Define L/D Max.

19. Where does the region of normal command lie on the total drag curve?

20. In the region of normal command, what must the pilot do to slow down while maintaining level flight?

21. Where is the region of reverse command on the total drag curve?

22. What must a pilot do in the region of reverse command to fly slower and why?

NOTES:

NOTES:

Chapter 5

High Lift Devices

Purpose of High Lift Devices:

→ To obtain a required lift force at lower velocities. (Does not increase the total lift capacity of a wing. The only way this can be done is by increasing the wing area.) This allows an aircraft to fly a slower approach speed and also to land at a slower speed.

→ Increase the value of C_LMax by:

1. Increasing the camber of an airfoil.
2. Delaying the boundary layer separation.

Camber Change:

→ The most commonly used method to increase C_LMax is by increasing the camber of the airfoil. Increasing camber increases C_LMax by creating a greater velocity differential between the top and bottom surfaces of the airfoil.

→ The most common method of changing a wing's camber is by the use of trailing edge flaps.

→ Increasing the camber increases C_LMax but decreases the stalling angle of attack.

(Reference Fig. 5-1)

Delay of Boundary Layer Separation:

(between wing and free airstream)

→ By increasing the energy level of the boundary layer over the wing, the wing can be rotated to a higher angle of attack before it stalls.

→ BLC (Boundary Layer Control) is used to increase the energy level of the boundary layer air. BLC increases C_L Max and the stalling angle of attack.

Types of BLC:

→Suction:

1. Suction BLC draws off the low energy, aerodynamically dead, turbulent air below the boundary layer, causing the high energy layers of air to be lowered closer to the airfoil. Suction BLC is inefficient due to the added weight of the equipment required to do the work. The increase in weight, therefore, partially offsets the advantages gained by the increased C_L Max.

→ Blowing BLC:

2. Blowing BLC increases the energy level of the boundary layer by introducing high energy air. It basically blows the turbulent air away from the top surface of the airfoil. It is more efficient and more commonly used than suction BLC. Jets get blowing, high energy air from engine compressor discharge air, therefore, there is not additional weight increase involved.

→ Vortex Generators:

3. Vortex Generators are small metal strips placed on the top of the wing usually in front of the control surfaces or towards the wing tips. The turbulence caused by these strips mixes high energy air from outside the boundary layer with boundary layer air.

Types of High Lift Devices:

✈ Trailing Edge Flaps: Most common type of high lift device. Increases camber and thus C_LMax. The disadvantages are when the flaps are put down there is a nose down pitching movement caused by the lift force moving towards the trailing edge of the wing. Flaps increase drag which during a landing approach allows for a steeper approach (better forward visibility) without increasing speed.

1. Plain and Split Flaps: the simplest types.
2. Fowler Flaps: the Fowler flap is the type of flap which is the most efficient aerodynamically. It moves rearward and down which increases the lift of the wing by increasing the camber (C_LMax) and wing area (S). The disadvantage is that it creates a large twisting moment so it must be used on extremely strong wings. It cannot be used on high speed, thin airfoils.
3. Slotted Flaps: Most flaps can be slotted to increase efficiency. Slotting adds BLC to camber change and the advantages of the two types of high lift devices are additive.

(Reference Fig. 5-2)

✈ Leading Edge Devices:

1. Leading Edge Flap: Increases C_LMax without causing a nose down pitching movement. (May actually pitch up.)
2. Leading Edge Slot: Fixed slot on the leading edge of the wing allows BLC at high angles of attack.
3. Leading Edge Slat: A leading edge slot that can be opened and closed. Increases camber, wing size and causes BLC.

→ Summary:

Devices that increase the area of the wing as well as the value of C_LMax are more efficient in decreasing stalling speed. If two high lift devices using differential principles are used, their effects are usually additive.

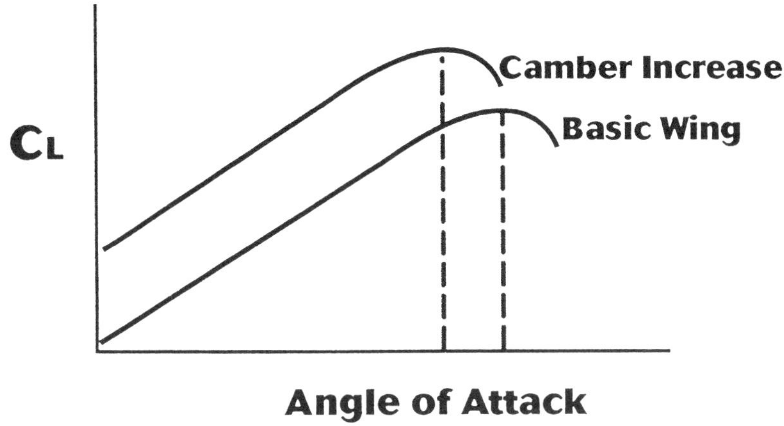

Figure 5-1

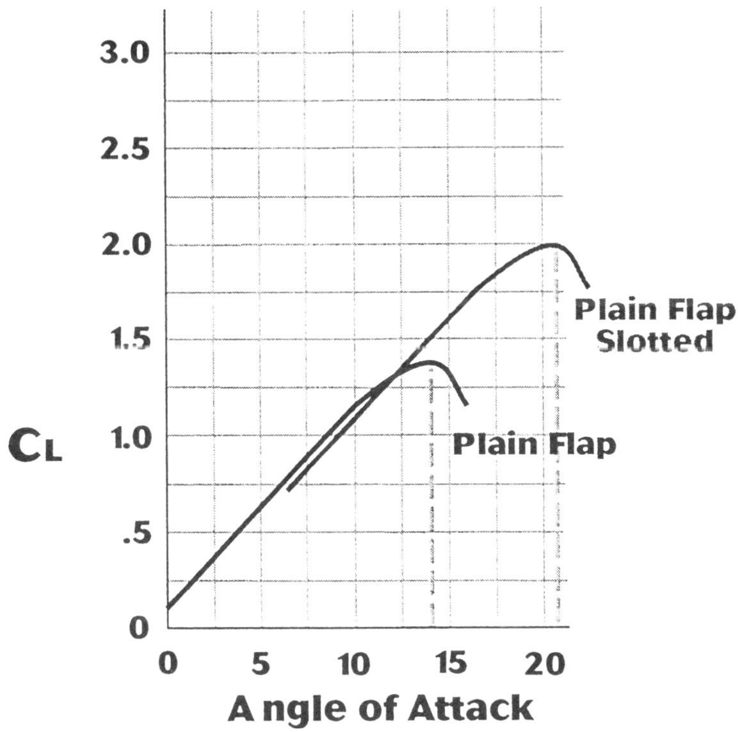

Figure 5-2

Study Question Chapter 5

1. What do high lift devices increase?

2. What are the two common ways high lift devices work?

3. What is the most common method of increasing C_LMax?

4. How does increasing the camber of a wing increase C_LMax?

5. What is the usual method of increasing camber?

6. What happens to the stalling angle of attack when C_LMax is increased by creating a positive camber?

7. What limits the maximum value of C_L?

8. How does boundary layer control increase the value of C_LMax?

9. What happens to the stalling angle of attack when C_LMax is increased by boundary layer control?

10. What are two types of BLC?

11. Describe suction BLC.

12. Why is suction BLC rather inefficient?

13. Describe blowing BLC.

14. Why is blowing BLC more efficient than suction BLC?

15. What are vortex generators?

16. What is the main advantage and disadvantage of using trailing edge flaps?

17. Describe the different types of flaps.

18. Why are flaps used for landing?

19. What is the advantage of using both types of high lift devices?

Chapter 6

Takeoff and Landing

Takeoff:

✈ General:

1. To get airborne, enough lift must be generated to balance weight.
2. Since an aircraft takes off from a standing start, an acceleration must be produced to get it airborne.
3. Thrust is the force that is most out of balance during takeoff.
4. Thrust must overcome drag and rolling friction during takeoff.
5. Therefore, the net accelerating force during takeoff is:

$$Fn = (T\text{-}D\text{-}Rf)$$

Fn = net accelerating force.
T = thrust.
D = drag.
Rf = rolling friction.

✈ Rolling Friction:

1. Rolling friction is the force acting parallel to the runway caused by the friction effect of the aircraft tires and the runway surface acted upon by the weight of the aircraft. Rolling friction is the greatest when the aircraft is motionless and decreases during the takeoff roll, until it becomes zero at takeoff.

✈ Factor Effecting Takeoff:

1. The only factor which effects lift during takeoff is dynamic pressure. When sufficient dynamic pressure has been generated for lift to equal weight, then the aircraft can fly.
2. Therefore, *takeoff IAS is only dependent upon the weight of the aircraft.*

✈ Takeoff Equation:

$$St = \frac{WV^2}{2g\,(T\text{-}D\text{-}Rf)}$$

St = takeoff distance.
W = weight.
$V^2 = TAS^2$
g = gravity.
(T-D-Rf) = net accelerating force.

✈ Factors Effecting Takeoff Distance:

1. Weight: Weight increases the amount of dynamic pressure required to produce lift equal to weight. Therefore, when q↑, then IAS↑, and TAS↑. An increase in lift to equal weight causes an increase in induced drag which causes total drag to increase. Therefore, the overall effect of a weight increase is an increase in takeoff distance.
2. Altitude/ Density: An increase in altitude decreases density. An increase in temperature decreases density. When density decreases, more TAS is needed to generate the required lift. Therefore, *any factor which decreases density (temperature increase, altitude increase, or humidity increase) will cause an increase in the required takeoff distance.*

An increase in altitude or temperature decreases the thrust available from a jet or prop and this causes an increase in the takeoff distance. The worst takeoff condition is : *High, Hot, Humid, and Heavy.*
3. Winds: a headwind decreases the takeoff distance because the amount of the headwind is a component of the required dynamic pressure. Conversely, a tailwind will greatly increase the required takeoff distances.

Landing:
✈ The same factors which effected the aircraft during takeoff also effect during landing. During landing, though, drag and rolling friction are aids to the pilot. Thrust is detrimental to stopping.

✈ Landing Equation:

$$Sl = \frac{WV^2}{2g(D+Rf-T)}$$

✈ Pilot techniques to reduce landing distances:
1. Aircraft should be as light as possible.
2. Landing should be made at the slowest possible safe speed.
3. Proper braking should be used.

✈ Types of Braking:
1. Aerodynamic Braking: is done by raising the nose of the aircraft after touchdown (not enough to drag the tail though!). This increases the angle of attack which increases induced and parasite drag. The increased drag helps slow the aircraft down. It is the most efficient type of braking.

2. Friction Braking: Most effective braking but can wear out mechanically and problems can arise such as hot brakes.

✈ Factors Effecting Landing Distance:
1. Weight: Increased weight causes an increased q (IAS) for lift to equal weight. This causes the landing TAS to be higher and this requires a greater amount of thrust. Therefore, an increases weight increases the landing distance.
2. Altitude/ Density: An altitude increase decreases density, therefore a higher TAS is required for the necessary IAS and this results in a longer landing distance.
3. Winds: A headwind decreases the landing distance because it becomes a component of the required dynamic pressure. Conversely, a tailwind will increase the landing distance.

Ground Effect:
✈ Caused by wing tip vortices striking the ground. Occurs within one wingspan distance from the ground. Decreases induced drag and thus decreases total drag.
✈ Landing: This causes an excess thrust condition which decreases the aircraft's rate of descent and causes the aircraft to float. The pilot must reduce thrust to overcome ground effect.
✈ Takeoff: When the aircraft gets out of ground effect, if it doesn't have an excess of thrust it could settle back onto the runway. Very possible when conditions are high, hot, humid, and heavy.

Study Questions Chapter 6

1. What are the five forces acting on an aircraft during takeoffs and landings?

2. Which force is most out of balance during the takeoff roll?

3. What two forces must thrust initially overcome on takeoff?

4. Define rolling friction.

5. When is rolling friction the greatest and when does it decrease?

6. What is the only factor which effects lift during takeoff?

7. What is net accelerating force?

8. What is the equation for takeoff distance?

9. Describe the effects of a weight increase on takeoffs.

10. What happens to density with an increase in altitude or temperature?

11. When density is decreased, what happens to IAS and TAS?

12. What is the overall effect on takeoff roll with a decrease in density?

13. What does an increase in altitude or temperature do to the thrust available from an engine?

14. How does a decrease in thrust effect takeoff distance?

15. What is the most dangerous takeoff condition?

16. Why are takeoffs into a headwind desirable?

17. What is the landing equation?

18. Discuss the effects of drag and rolling friction on landings.

19. Discuss the effects of a weight reduction on landings.

20. What speed should landings be made at?

21. What are the two types of braking and describe them?

22. Which type of braking is the most efficient?

23. Which type of braking is the most effective?

24. Discuss the effects of a weight increase on landings.

25. Discuss what happens to an aircraft landing at a high altitude airport.

26. What happens to landing distance when landing in conditions of high temperature?

27. What causes ground effect?

28. What effect does ground effect have on induced and total drag when within one wingspan of the ground?

29. Why does an aircraft tend to float when it gets into ground effect?

30. What corrections must a pilot make when he gets into ground effect?

31. What effect does ground effect have on takeoffs?

Chapter 7

Spins

Definitions:
A spin is an aggravated stall which results in autorotation. Autorotation is a combination of roll and yaw.

Three Motions of a Spin:
✈ Descent: Descent results from the fact that since the aircraft is in a stalled condition there isn't enough lift generated to balance the weight.
✈ Roll: Roll is initiated by the aggravation of the stall and the decrease in the angle of attack on the upgoing wing.
✈ Yaw: Yaw is caused by the increase in the angle of attack on the downgoing wing.

Effect on the Individual Wings:
✈ It must be remembered that the entire spin evolution takes place in a stalled condition (at angles of attack higher than the stalling angle of attack).
✈ To see what happens to each wing during a spin, plot C_L and C_D against angle of attack on the graph. (Reference Fig. 7-1)

✈ Upgoing Wing: Angle of attack ↓, L↑, D↓. This causes the upgoing wing to roll more in the direction of the spin.

✈ Downgoing Wing: Angle of attack ↑, L↓, D↑. The increase in drag causes the yaw towards the downgoing wing. The combination of the increased roll on the upgoing wing and the increased yaw on the downgoing wing causes the autorotation and it will continue until some action stops it.

Spin Recovery:

✈ Stop the Autorotation: This is accomplished by forcing the nose of the aircraft back into the relative wind. It is usually done by applying rudder opposite to the direction of the spin. When both wings have the same angle of attack the autorotation motion will stop.

✈ Reduce the Angle of Attack: Once the autorotation stops, simply recover from the stall by reducing the angle of attack.

✈ Usual Steps for Spin Recovery in Airplanes:

1. Reduce power: because of the steep descent.
2. Opposite rudder: stop the autorotation.
3. Forward yoke/ stick: to reduce the angle of attack.

Phases of a Spin:

✈ Incipient Spin: It is that portion of a spin from the time the aircraft stalls and rotation starts until the spin is fully developed.

✈ Fully Developed Spin: A fully developed spin means that the angular rotation rates, airspeed, and vertical speed are stabilized from turn to turn and the flight path is close to vertical.

Types of Spins:

✈ Erect Spins: +1 "G" and a maximum unit angle of attack.

✈ Inverted Spins: -1 "G" and a 0 unit angle of attack.

✈ Flat Spins: The flat spin has a near level pitch and roll attitude with the spin axis near the CG of the aircraft. Recovery from a flat spin may be extremely difficult or even impossible because the relative wind is directly under the aircraft so the control surfaces do not have any effect.

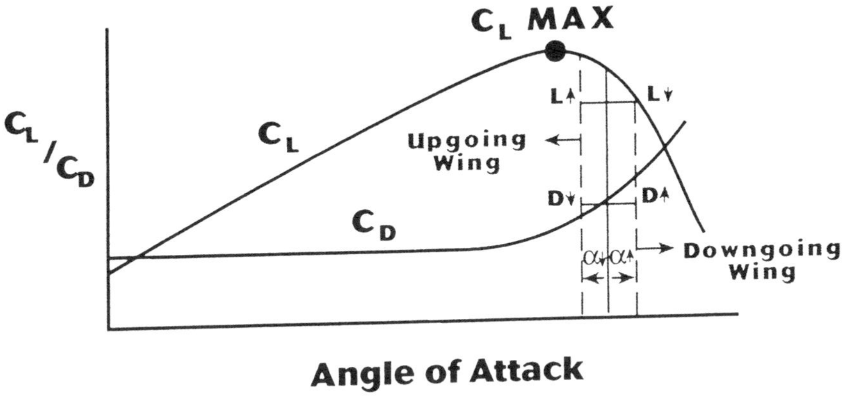

C_L MAX
C_L / C_D
C_L
C_D
L↑
L↓
Upgoing Wing
D↓
D↑
α↓
α↑
Downgoing Wing
Angle of Attack

Figure 7-1

Study Questions Chapter 7

1. Define a spin.

2. What are the three distinct motions of a spin?

3. Describe what happens to the upgoing wing during a spin.

4. Describe what happens to the downgoing wing during a spin.

5. What is the first objective in spin recovery?

6. How do you stop autorotation?

7. What is the second step in recovering from a spin?

NOTES:

NOTES:

Chapter 8

VG/VN Diagram

Strength of Metals:

✈ Stress: stress is the force per unit area.

✈ Strain: strain is the unit deformation.

✈ Elastic Limit: Elastic limit or proportional limit is the point up to which if a metal is stressed, it will return to its original shape. Stress greater than the elastic limit will cause strain or permanent deformation as the metal will not return to its original shape.

✈ Ultimate Stress: It is the maximum stress before a material breaks.

Vg/ Vn Diagram:

✈ General Graph Outline:

1. The Vg/ Vn diagram displays the operating strength limitations of an aircraft in graph form. It becomes the "flight envelope".
(Reference Fig. 8-1)

2. Axis of the Graph: The vertical axis is load factor (L/W = N or G). The horizontal axis is dynamic pressure (IAS).

3. Redline Airspeed (Vne): It is the highest

permissible airspeed (IAS) before excessive dynamic pressures can cause control problems and/ or aircraft damage.

✈ Phenomenon of Flight Above Redline Airspeed:

1. Wing Flutter: During wing flutter the wing and control surfaces tend to flutter up and down and this can cause them to fail.
2. Wing Divergence: In wing divergence the wing fails immediately due to a change in the lift force which produces a twisting moment which increases with lift.
3. Aileron Reversal: An aileron input causes a bending of the wing's leading edge. This alters the wing's angle of attack and the aircraft rolls in the opposite direction than desired.
4. Shock Wave Formation: Shock waves can form on the aircraft and can cause structural damage due to the compressibility effects of the air.

✈ Vg/ Vn Specifics:

1. Accelerated Stall Line: The accelerated stall line divides the graph between the safe flight region and the stall area. It is the largest load factor that can be generated for a given IAS before a stall occurs. An aircraft can take greater positive than negative g's. The accelerated stall line is actually a line of C_LMax's.
2. Limit Load or Normal "G" Load: Limit load is the maximum load factor that an aircraft can sustain during operation without causing an over stress condition. It is the elastic limit.

3. Ultimate Load: The ultimate load is the greatest load factor that the aircraft can sustain without structural failure. It is 1.5 times the limit load.
4. Maneuvering Speed (Va): Maneuvering speed is the lowest speed at which the aircraft can develop a load factor equal to the limit load, therefore, Va occurs at the intersection of the accelerated stall line and the limit load. Any speed above Va could cause an overstress, therefore, it is the highest speed to fly in turbulence. Maneuvering speed decreases as the aircraft's weight decreases because the aircraft stalls at slower speeds and this shifts the accelerated stall line to the left on the graph.

✈ Factors Effecting the Safe Flight Envelope:
1. Gross Weight: If the aircraft's gross weight increases, the amount of load factor that the wings can generate decreases. This causes the limit and ultimate loads to decrease.
2. Altitude: An increase in altitude reduces the aircraft's ability to generate load factor. An increase in altitude causes the redline airspeed to decrease (supersonic flight only).
3. Landing Configuration: The entire safe flight envelope shrinks in all dimensions due to the lower maximum landing gear and flap speeds.
4. Turbulence: Fly at maneuvering speed or less for your aircraft weight because by doing so the aircraft will stall before an over stress condition is encountered.

Aircraft Categories Based on Load Factor: (Vg/ Vn)

✈ Normal Category: A normal category aircraft (most trainers) has a maximum certificated limit load of +3.8 G's. Normal category aircraft may perform the following maneuvers:

1. Any maneuver incident to normal flying.
2. Stalls.
3. Lazy eights, chandelles, and steep turns not in excess of a 60 degree angle of bank.

✈ Utility Category: A utility category aircraft has a maximum certificated limit load of +4.4 G's. Utility aircraft may perform the following maneuvers:

1. All maneuvers approved for a normal category aircraft.
2. Spins (if approved for that aircraft).
3. Lazy eights, chandelles, and steep turns with angles of bank greater than 60 degrees.

✈ Acrobatic Category: An acrobatic category aircraft has a maximum certificated limit load of +6.0 G's and can perform any maneuver except those which are specifically prohibited for that aircraft.

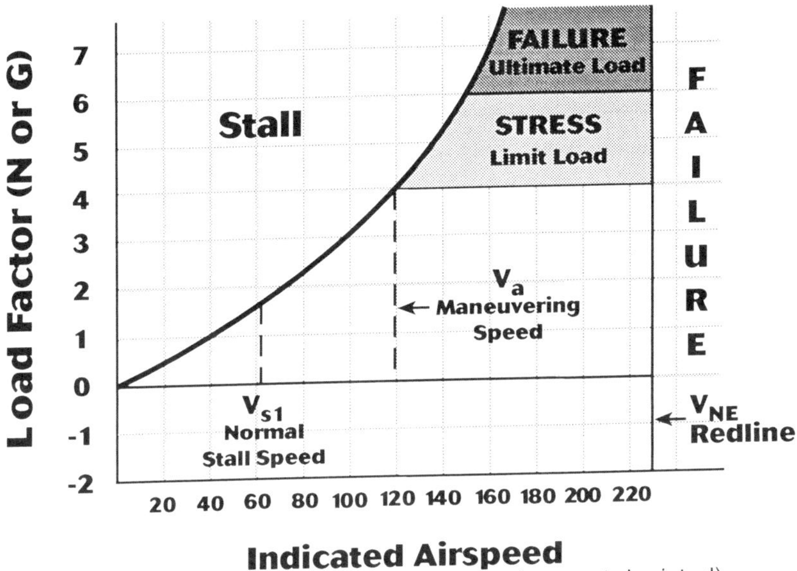

(Note: Negative load factor flight envelope not depicted)

Figure 8-1

Study Questions Chapter 8

1. What does the Vg/ Vn diagram outline?

2. What are the two axis of the Vg/ Vn diagram?

3. What is the vertical line at the right side of the Vg/ Vn diagram?

4. What can occur above red line speed?

5. Define wing divergence and where it occurs.

6. Where does wing flutter occur?

7. What happens in aileron reversal?

8. Above what speed will aileron reversal occur?

9. What is the main purpose of the red line speed?

10. What can occur at any G load greater than the limit load?

11. What can occur at the ultimate load?

12. How much greater is the ultimate load than the limit load?

13. What is the accelerated stall line?

14. Can over stress occur if the aircraft is flown at velocities up until the limit load is reached?

15. Define maneuvering speed.

16. What can happen at any velocity greater than maneuvering speed?

17. How does maneuvering speed change with a decrease in an aircraft's weight?

18. What happens to the size of the Vg/ Vn diagram in the landing configuration?

19. What is the certificated positive limit load of a normal category aircraft? a utility category aircraft? and an acrobatic category aircraft?

Chapter 9

Stability

Definition:

✈ Stability is the property that causes an object when disturbed from equilibrium to develop forces or moments which tend to return that object to equilibrium.

Classification of Stability:

✈ Positive: When displaced from an equilibrium position, the object will eventually return to the equilibrium position.

✈ Negative: When displaced from an equilibrium position, the object will move farther away from the original equilibrium position (instability).

✈ Neutral: When displaced from an equilibrium position, the object will simply come to rest at that new position.

Further Classifications of Stability:

✈ Static: Static stability is the tendency an object possesses after being displaced from equilibrium.

✈ Dynamic: Dynamic stability is the motion created after an object is displaced from equilibrium.

Types of Stability:

✈ Longitudinal: Longitudinal stability is the motion tendency of the longitudinal axis about the lateral axis (pitch).

✈ Directional: Directional stability is the motion tendency of the longitudinal axis about the vertical axis (yaw).

✈ Lateral: Lateral stability is the motion tendency of the lateral axis about the longitudinal axis (roll).

Airfoils that Effect Stability:

✈
1. Fuselage
2. Wing
3. Horizontal Stabilizer
4. Vertical Stabilizer
5. Location of the aerodynamic center (AC) in relationship to the center of gravity(CG).

Longitudinal Stability:

✈ Longitudinal stability is the motion tendency of the longitudinal axis about the lateral axis (pitch).

Factors Effecting Longitudinal Stability:

✈ Wing: The AC is in front of the CG which causes a negative contributor to longitudinal stability. The AC is the point through which all of the lift on the wing is considered to be acting. This causes a large moment arm and when a force pitches the aircraft's nose up, the large moment arm between the AC and CG causes the nose to move away from equilibrium.
(Reference Fig. 9-1)

→ Fuselage: The AC is in front of the CG so the fuselage as a negative contributor like the wing.

→ Horizontal Stabilizer: The horizontal stabilizer is the largest positive contributor to longitudinal stability. It is symmetrical so lift is dependent on angle of attack. When the nose of the aircraft pitches up, the horizontal stabilizer acquires a positive angle of attack which allows it to generate lift. This upward lift vector acting at such a long distance from the CG causes the nose to pitch down which cancels out the nose up motion caused by the wing and fuselage.
(Reference Fig. 9-2).

→ Vertical Stabilizer: No effect.

→ Sweepback Wing: The sweepback design of a wing forces the AC aft towards the CG which reduces the moment arm and causes a positive contributor to longitudinal stability. Aircraft with sweepback wings can have glove vanes which come out to move the AC forward for more maneuverability when needed. The F-14 uses this concept. (Concept to remember is that stability and maneuverability are at opposite ends of the spectrum. For an aircraft such as a jet fighter to be maneuverable it gives up a lot of stability and conversely a civilian trainer aircraft gives up a lot in maneuverability to be a stable teaching platform.)

→ Aircraft Without a Horizontal Stabilizer: The most common design is to sweep back the wings or mount the wing further aft until the AC of the wing is aft of the CG. This creates a sufficient moment arm which overcomes the negative effect of the fuselage.

✈ Thrust Axis: When the thrust axis coincides with the longitudinal axis it is a neutral contributor. When the thrust axis is above the CG and power is added the aircraft will pitch down. When the thrust axis is below the CG and power is added the aircraft will pitch up.

Directional Stability:

✈ Directional stability is the motion of the longitudinal axis about the vertical axis (yaw).

✈ Main Factor Causing Directional Stability Problems: A side slip condition is the main factor effecting directional stability. A side slip is defined as an angle between the relative wind and the longitudinal axis.

✈ Factors Effecting Directional Stability:

1. Fuselage: The AC is forward of the CG, therefore, the effects of the fuselage tend to increase the side slip so the fuselage is a negative contributor to directional stability.
2. Wing: When a side slip is produced, the forward moving wing is advancing into the relative wind faster than the backward moving wing. Dynamic pressure is increased on the forward wing and decreased on the other. This causes a drag differential which tends to pull the advancing wing back so the wing is a positive contributor to directional stability. The wing's positive contribution is very weak, so conventional aircraft use a vertical stabilizer to gain directional stability.

3. Horizontal Stabilizer: No effect.
4. Vertical Stabilizer: The vertical stabilizer is the greatest positive contributor to directional stability. A side slip causes a sideways angle of attack to occur on the vertical stabilizer. Since its AC is aft of the CG, the sideways lift vector of the vertical stabilizer tends to return the aircraft to equilibrium. The shorter the fuselage, the shorter the moment arm will be and thus a larger vertical stabilizer will be required to obtain directional stability.
5. Sweepback: Sweepback creates more lift on the forward moving wing during a side slip, therefore, it also creates more drag which pulls the fuselage back towards equilibrium.
6. Asymmetrical Thrust: If one engine of a multi-engine aircraft is not working this creates a thrust moment arm which is a negative contributor to directional stability.

Lateral Stability:

✈ Lateral stability is the motion tendency of the lateral axis about the longitudinal axis (roll).

✈ Factors Effecting Lateral Stability:
1. Fuselage: No effect.
2. Horizontal and Vertical Stabilizers: No effect.
3. Wing: The wing is a major contributor to lateral stability. The upgoing wing angle of attack decreases, with lift decreased. The downgoing wing angle of attack increases with lift also increased. The increased lift vector on the downgoing wing tends to raise the wing back to equilibrium.

4. High Wing: The AC is above the CG. In banked conditions this creates a moment arm and force that tends to return the wings to the level equilibrium state. Since the vertical component of lift is laterally displaced it is the larger weight moment arm that returns the wings to level. Because of this a high wing is a positive contributor to lateral stability. (High wing aircraft make good trainer or cargo aircraft due to their positive lateral stability.)
(Reference Fig. 9-3)

5. Low Wing: A low wing is a negative contributor to lateral stability because the moment arm and force that are created when it is moved from equilibrium tend to continue moving the wing further out of the equilibrium state. (A low wing aircraft has excellent maneuverability and thus many fighter aircraft are low winged.)
(Reference Fig. 9-4)

6. Dihedral: Dihedral is a positive angle between the wing and the lateral axis. When displaced from equilibrium the following effects occur: The upgoing wing angle of attack decreases, and the lift decreases. The downgoing wing angle of attack increases, and the lift also increases. The increased lift on the downgoing wing then tries to return the wing to equilibrium. Dihedral is positive contributor to lateral stability. It is also how some lateral stability can be built into a low winged aircraft.
(Reference Fig. 9-5)

7. Anhedral: Anhedral is negative dihedral and is a negative contributor to lateral stability because the upgoing wing gains additional lift when displaced from equilibrium which tends to lift the wing further from the equilibrium state. Anhedral is how maneuverability can be built into a high wing aircraft. An example of this is the Marine Corps' Harrier. (Reference Fig. 9-6)

8. Sweepback: Sweepback is a positive contributor to lateral stability because during a turn the inside wing has more chordwise flow. This increases its lift and tends to return the wing to its original equilibrium condition.

✈ Most Stable Wing: The most stable wing in terms of lateral stability is a high mounted, sweepback wing with dihedral.

Cross Effects:

✈ Adverse Yaw: Adverse yaw is yaw produced as a result of rolling the aircraft. It is caused by drag and lift differential. The aircraft yaws in the opposite direction of the roll due to the increased drag on the upgoing wing. The purpose of the rudder is to overcome adverse yaw and that is why it is so important to fly coordinated turns.

✈ Proverse Roll: Proverse roll is roll produced by yawing the aircraft. It is caused by the lift differential and spanwise flow. The aircraft rolls in the direction of the yaw because of the increased lift on the forward moving wing. (This can be easily demonstrated by turning an aircraft by using the rudder only.)

✈ Spiral Divergence: Spiral divergence is yaw coupled with roll in the same direction. It is caused by strong directional stability and weak lateral stability. (Most fighter and attack aircraft have this.)

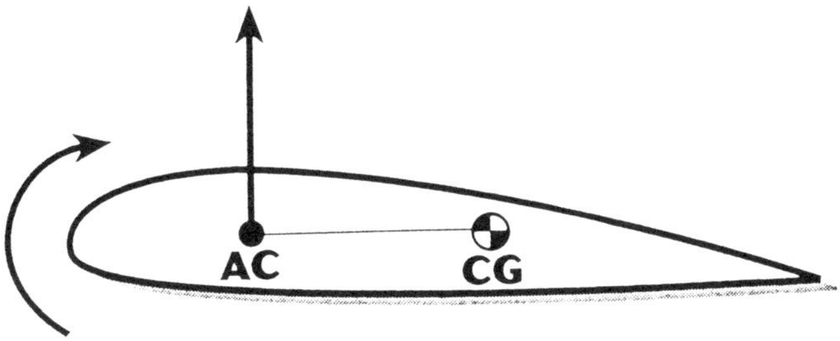

Figure 9-1

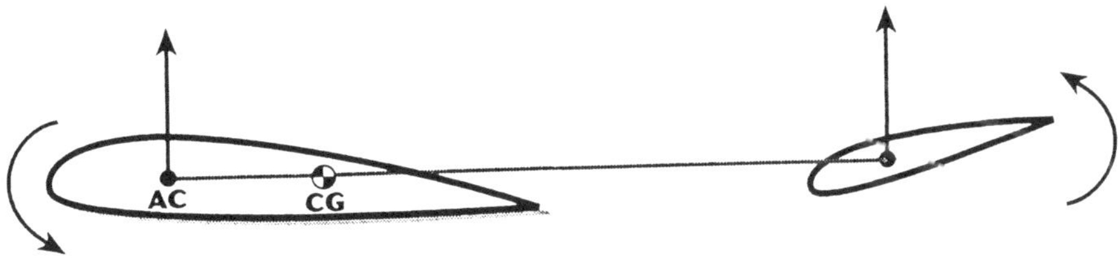

Figure 9-2

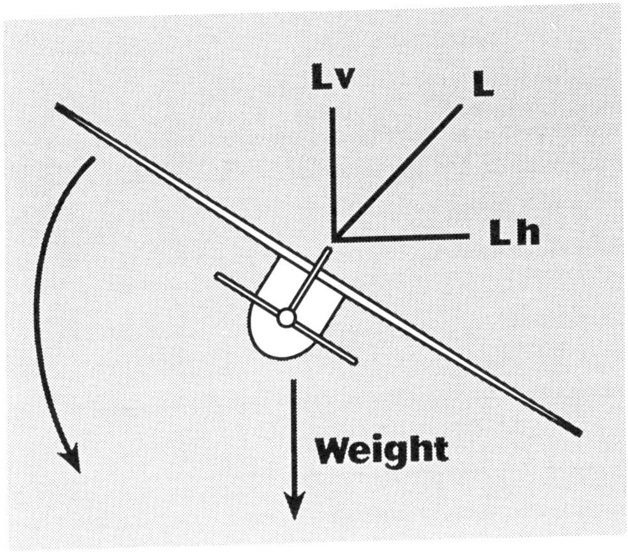

Figure 9-3

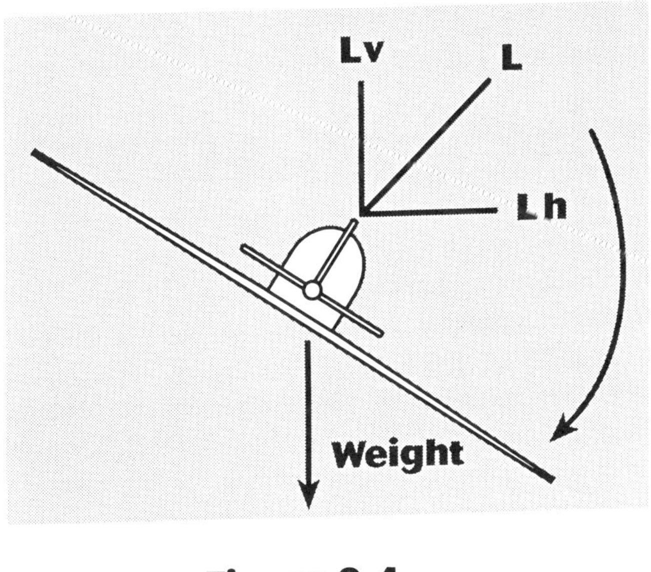

Figure 9-4

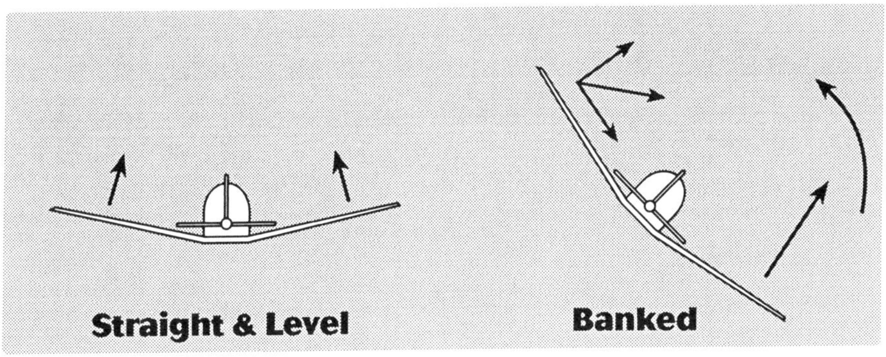

Figure 9-5

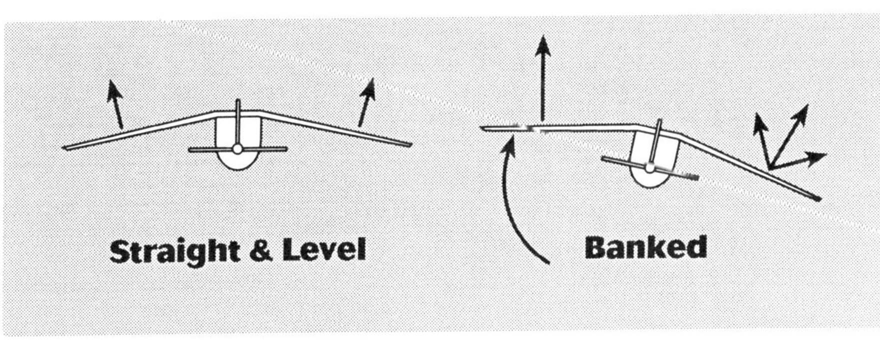

Figure 9-6

Study Questions Chapter 9

Questions 1-9 define the following terms:

1. Stability.

2. Dynamic stability.

3. Static stability.

4. Positive stability.

5. Negative stability.

6. Neutral stability.

7. Longitudinal stability.

8. Directional stability.

9. Lateral stability.

10. Is the location of the aerodynamic center of a wing with respect to the aircraft's center of gravity very important to longitudinal stability?

11. On conventional aircraft, where is the AC in relation the CG?

12. What type of influence does a conventional wing have on longitudinal stability?

13. On aircraft designed with a conventional wing, what is needed to compensate for the wings negative influence?

14. Where is the AC of a fuselage generally located in relation to the CG?

15. What type of contributor is the fuselage to longitudinal stability?

16. What is needed to balance the negative effect of the fuselage?

17. Describe how the horizontal stabilizer overcomes the negative effect of the wing and fuselage.

18. What must be done to the AC of the wing on aircraft designed without a horizontal stabilizer?

19. Why is the thrust axis normally aligned with the longitudinal axis?

20. What effect does sweepback have on longitudinal stability?

21. What happens to the pitch maneuverability of aircraft that sweep their wings and have a horizontal stabilizer?

22. What are glove vanes used for?

23. What is the main factor effecting directional stability?

24. Define side slip.

25. What type of contributor is the fuselage to directional stability?

26. What type of contributor is the wing to directional stability and why?

27. Why do conventional aircraft utilize vertical stabilizers?

28. What contributing factor is the vertical stabilizer to directional stability?

29. What basically determines the size of the vertical stabilizer?

30. What factor is assymetrical thrust to directional stability?

31. How does the wing create positive lateral stability?

32. What type of lateral stability does a high mounted wing have? a low mounted wing?

33. How does dihedral effect lateral stability?

34. How does anhedreal effect lateral stability?

35. How does wing sweepback effect lateral stability?

36. What is the most stable aircraft wing design?

37. How are maneuverability an stability related?

38. Describe adverse yaw.

39. Describe proverse roll.

40. What causes spiral divergence?

Chapter 10

Aircraft Performance

Thrust and Thrust Horsepower curves:

✈ Thrust Required (T_R): In equilibrium flight thrust equals drag. Therefore, in equilibrium flight the total drag curve will also be the thrust required (T_R) curve. Thrust required is the force needed to overcome the existing drag force to maintain equilibrium. If thrust doesn't equal drag, the aircraft will either speed up or slow down in level flight.
(Reference Fig. 10-1)

✈ Thrust Available (T_A): Thrust available is the thrust that is actually available from either a propeller driven aircraft or a jet.
1. Propeller (T_A):Thrust available (efficiency) from a propeller decreases as velocity increases.
(Reference Fig. 10-2)

2. Jet (T_A): Thrust available for a jet aircraft is basically a straight line because a jet will draw in all the air it needs to produce the necessary thrust at any airspeed.
(Reference Fig. 10-3)

✈ For both props and jets, an increase in altitude decreases density and therefore also decreases thrust available.

✈ Plotting T_R and T_A Together:
The difference between T_A and T_R is excess thrust (T_E). A positive value means that there is more thrust available than drag so acceleration can occur. At V1 on the graphs, $T_A = T_R$, therefore, V1 is the maximum airspeed for straight and level flight. Any velocity greater than V1 there will be more drag than thrust available and the aircraft would slow down. The T_A curve represents the maximum fuel flow for a jet. The minimum fuel flow for a jet occurs at the airspeed for L/ D Max.
(Reference Fig. 10-4)

✈ Thrust Horsepower (THP):
1. THP Required (THP_R):
Propeller driven aircraft must convert fuel flow into horsepower before it is converted to thrust. Fuel flow for a prop then is obtained by developing a THP curve. The THP_R curve is derived from the thrust curve by the formula: $THP = TV/325$. THP_R for a jet can be computed in the same manner and will look the same.
(Reference Fig. 10-5)

2. THP Available (THP_A):
THP_A for a prop decreases with an increase in airspeed at higher speeds because the propeller becomes less efficient.
(Reference Fig. 10-6)

3. Locating L/ D Max on THP Curve:
In order to locate L/ D Max on the THP curve the thrust curve (T_R) must be used. It is found by drawing a horizontal line parallel to the velocity axis and touching the bottom of the drag/ thrust required curve at L/ D Max.
(Reference Fig. 10-7)
This line can be transferred to the THP_R curve by the THP = TV/ 325 relationship. It then becomes a straight slanted line starting at the origin and going tangent to the THP_R curve. Tangent point is L/ D Max.
(Reference Fig. 10-8)

Performance Parameters:

✈ Maximum Angle of Climb (Vx): The maximum angle of climb provides for the maximum vertical distance for the least horizontal distance. It is the airspeed used for an obstacle clearance takeoff. The airspeed for the maximum angle of climb occurs where there is the greatest excess thrust. Maximum angle of climb airspeed for a jet is at L/ D Max. Maximum angle of climb airspeed for a prop is less than L/ D Max.
(Reference Fig. 10-9)

✈ Maximum Rate of Climb (Vy): Airspeed for the maximum rate of climb gives the maximum vertical distance in the least amount of time. It enables an aircraft to arrive at its operating altitude in the shortest time. Want to do as much work in as short a period of time as possible, therefore, want the maximum excess power available.

$$THP_E = THP_A - THP_R$$

The airspeed for the maximum rate of climb for a jet is greater than L/ D Max. The airspeed for the maximum rate of climb for a prop is in the vicinity of L/ D Max. *Maximum rate of climb airspeed is faster than the airspeed for maximum angle of climb.*
(Reference Fig. 10-10)

✈ Maximum Endurance: To fly at maximum endurance airspeed is to stay airborne for the maximum amount of time. It will occur at the airspeed derived from the minimum fuel flow.
1. A jet has a direct relationship between fuel flow and thrust. Minimum fuel flow (maximum endurance airspeed) occurs at the airspeed for L/ D Max.
(Reference Fig. 10-11)

2. A prop aircraft has a direct relationship between fuel flow and horsepower. The bottom of the THP_R curve (maximum endurance airspeed) is less than the airspeed for L/ D Max.
(Reference Fig. 10-12)

✈ Maximum Range: To fly as far as possible with the present fuel supply, the airspeed for maximum range will be chosen. It occurs at the best relationship between fuel flow and velocity, therefore, it maximizes fuel flow and velocity.
1. Jet: Maximum range for a jet is on the thrust curve. It is the point where you get the greatest increase in velocity for an increase in fuel flow. Maximum range airspeed for a jet is greater than the airspeed for L/ D Max. It can also be found by drawing a line tangent

to the thrust curve (T_R).
(Reference Fig, 10-13 and Fig. 10-14)

2. Prop: A prop aircraft uses the THP_R curve because it deals with fuel flow. Maximum range is found by drawing a line tangent from the origin to the THP_R curve. Maximum range airspeed for a prop driven aircraft occurs at the airspeed for L/ D Max.
(Reference Fig. 10-15)

✈ Maximum Glide Endurance: You fly at the airspeed for maximum glide endurance when you want to stay up as long as possible after an engine quits, so you want a rate of descent as small as possible. Maximum glide endurance is concerned with the maximum time to get to the ground and not the distance over the ground. You want the wing to do as little work as possible over time:

$$THP = work/ time = FD/ time = TV/ 325$$

Therefore, the airspeed for maximum glide endurance occurs at the bottom of the THP_R curve. Maximum glide endurance airspeed occurs at less than L/ D Max airspeed.
(Reference Fig. 10-16)

✈ Maximum Glide Range:
Since you want to glide the farthest distance you want the airspeed that equates to the maximum lift for the minimum drag. Maximum glide range airspeed is found on the T_R curve. Maximum glide range airspeed is equal to L/ D Max airspeed for a jet and a prop.
(Reference Fig. 10-17)

Effects of Weight, Altitude, and Headwind on Performance:

✈ Weight: An increase in weight decreases all parameters except for glide range. A heavier aircraft will come down faster, but will go the same distance.

✈ Altitude Increase (causes an increase in power required and a decrease in power available):

1. Jet: Maximum angle of climb decreases because Vx speed increases. Maximum rate of climb decreases because Vy speed decreases. Maximum endurance increases because less fuel is used. Maximum range increases. Maximum glide endurance increases. Maximum glide range increases.

2. Prop: Maximum angle of climb decreases because Vx speed increases. Maximum rate of climb decreases because Vy speed decreases. Maximum endurance decreases for a given brake horsepower (BHP), the prop burns more fuel at altitude. Maximum range has no change (uses more fuel but goes faster). Maximum glide endurance increases. Maximum glide range increases.

✈ Headwind: Jet/ Prop: No effect on endurances. Decreases ranges. No effect on maximum rate of climb (still gets to altitude in the same time, but covers less distance over the ground). Increases maximum angle of climb because distance traveled over the ground is less.

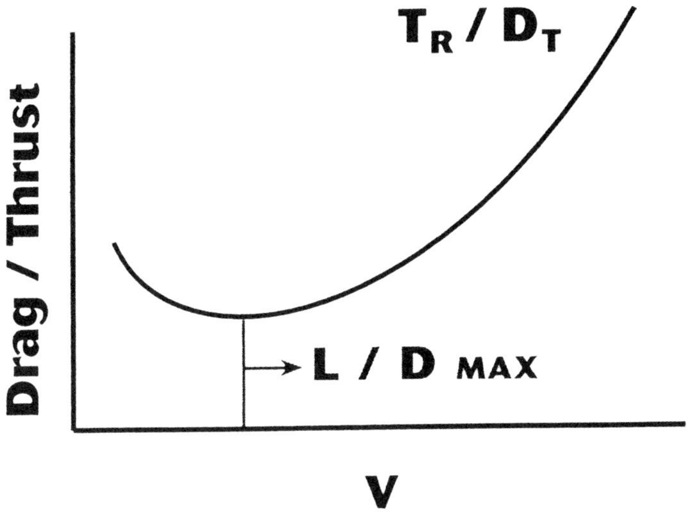

Figure 10-1

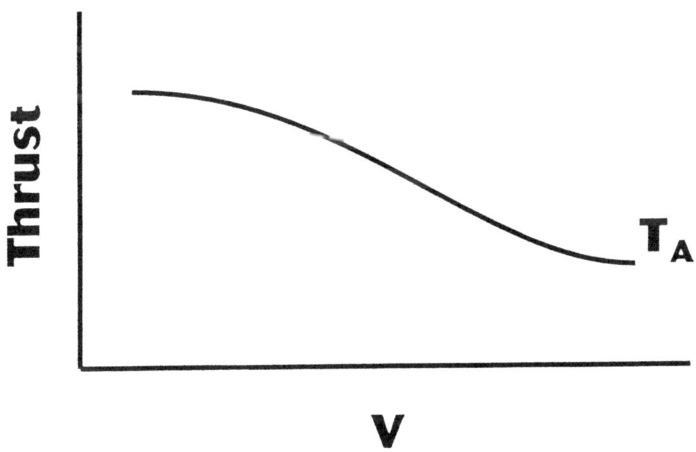

Figure 10-2

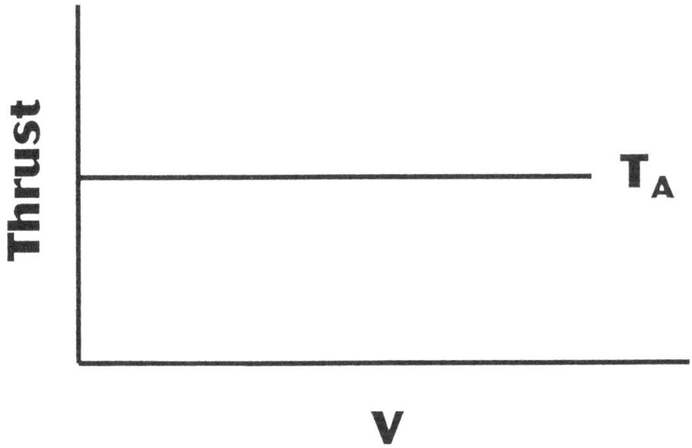

Figure 10-3

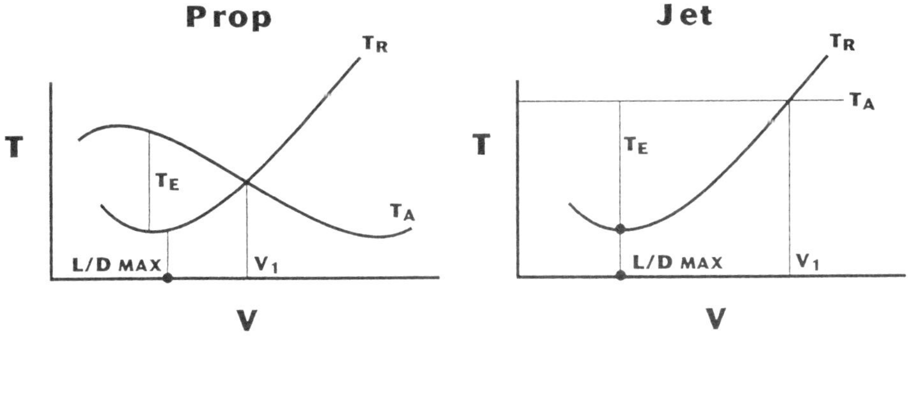

Figure 10-4

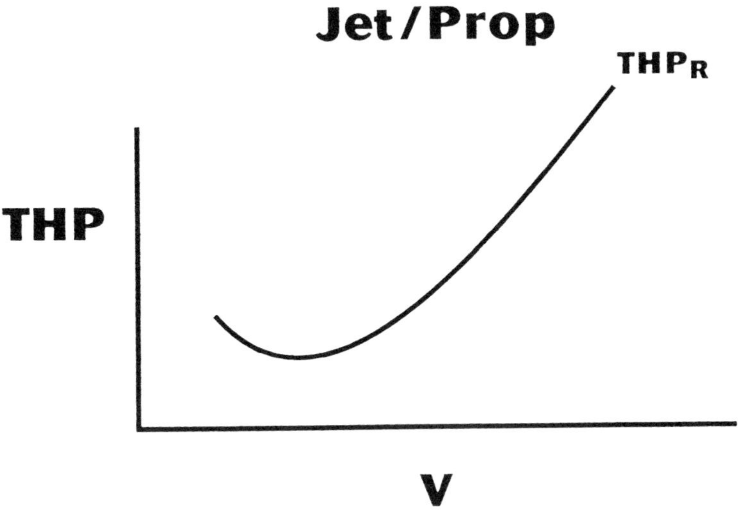

Figure 10-5

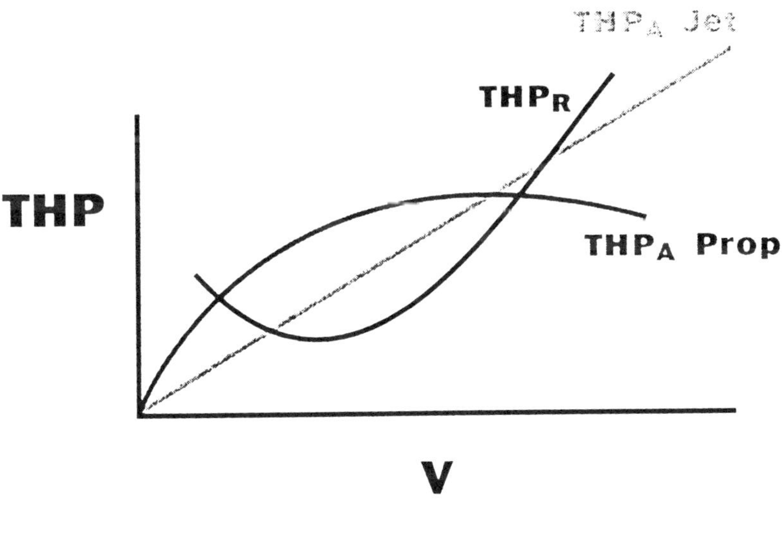

Figure 10-6

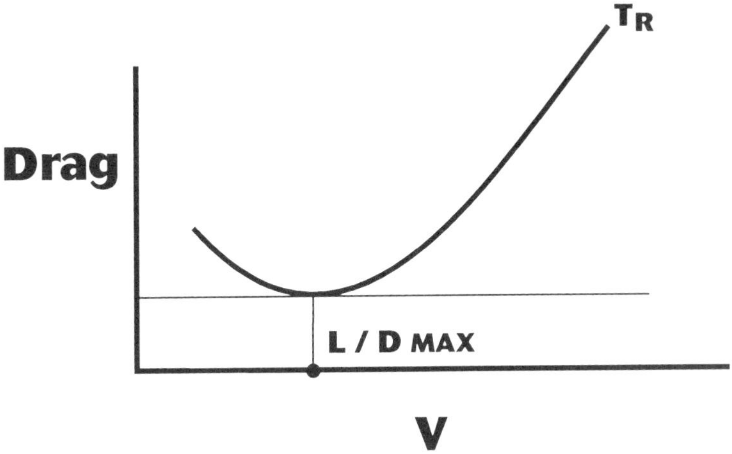

Figure 10-7

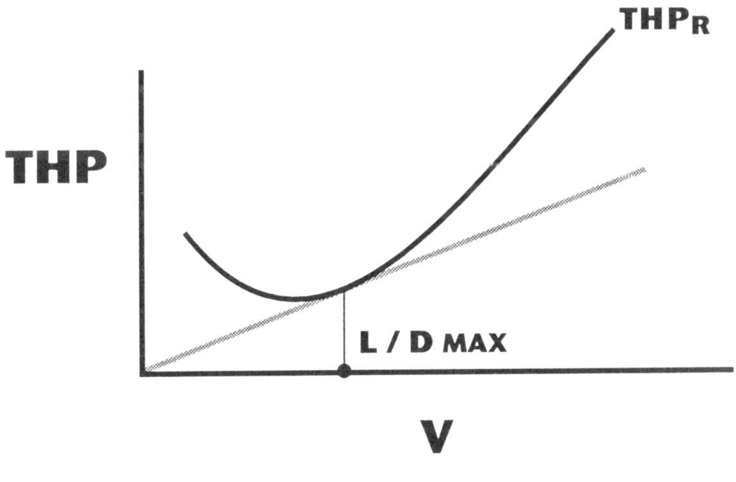

Figure 10-8

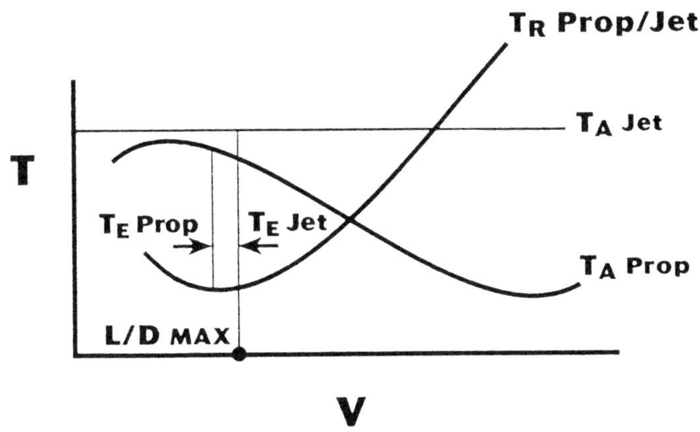

Figure 10-9

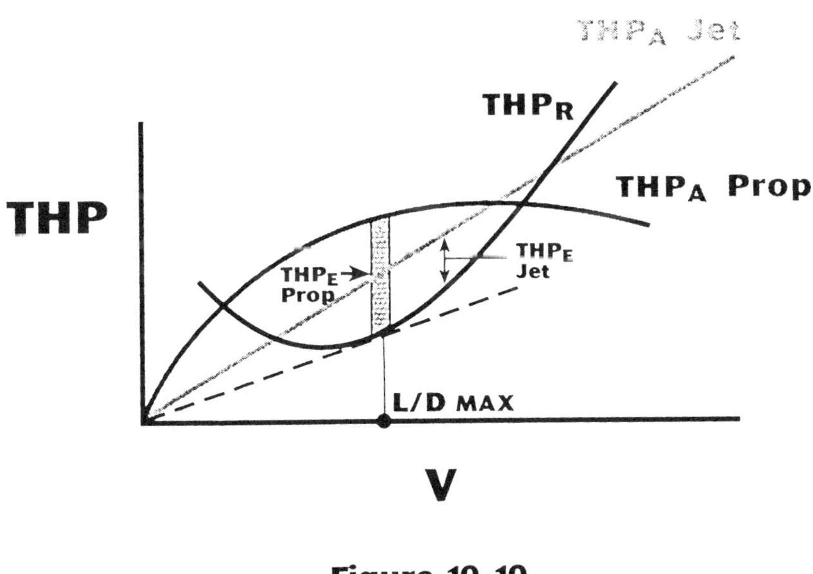

Figure 10-10

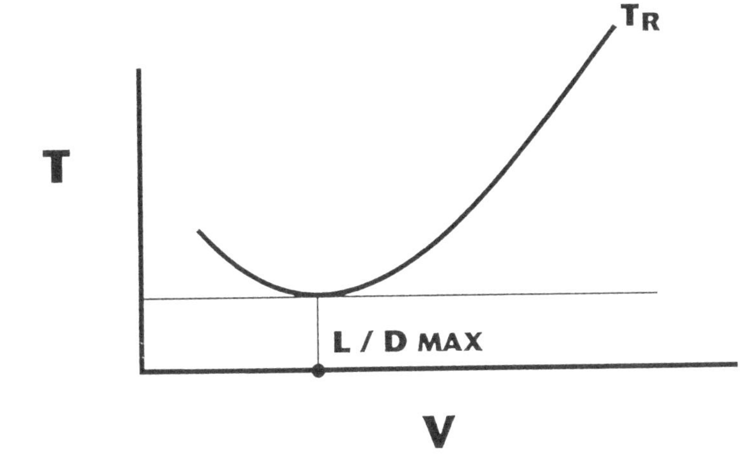

Figure 10-11

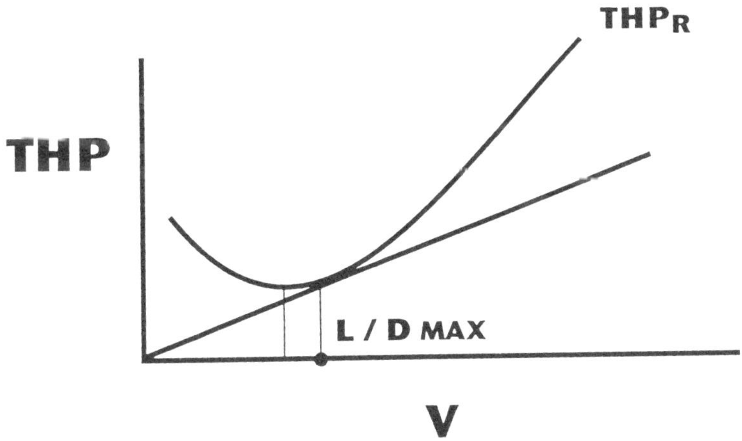

Figure 10-12

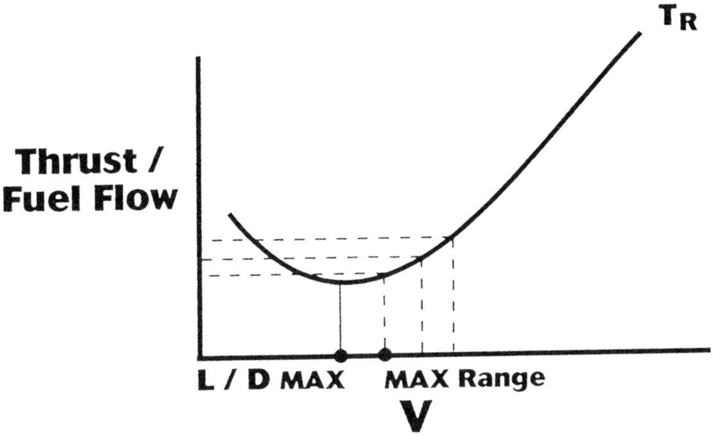

Figure 10-13

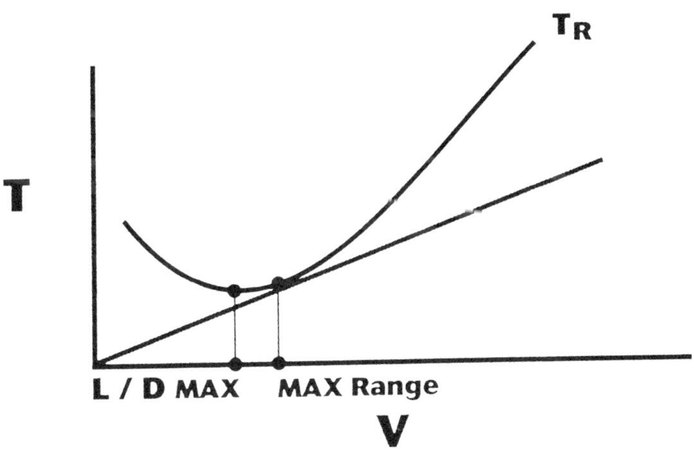

Figure 10-14

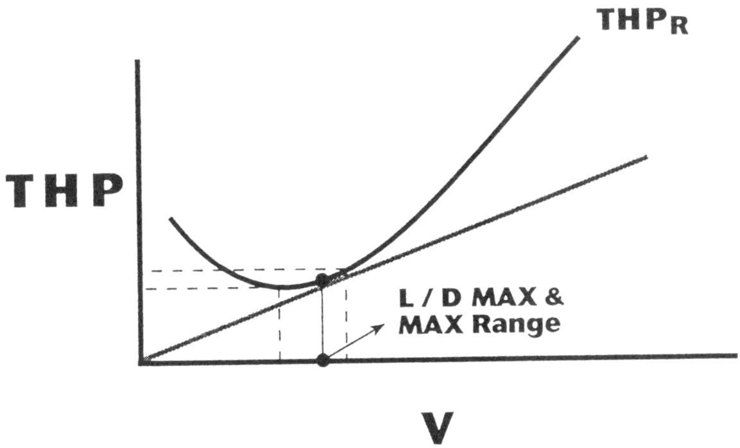

Figure 10-15

Prop / Jet

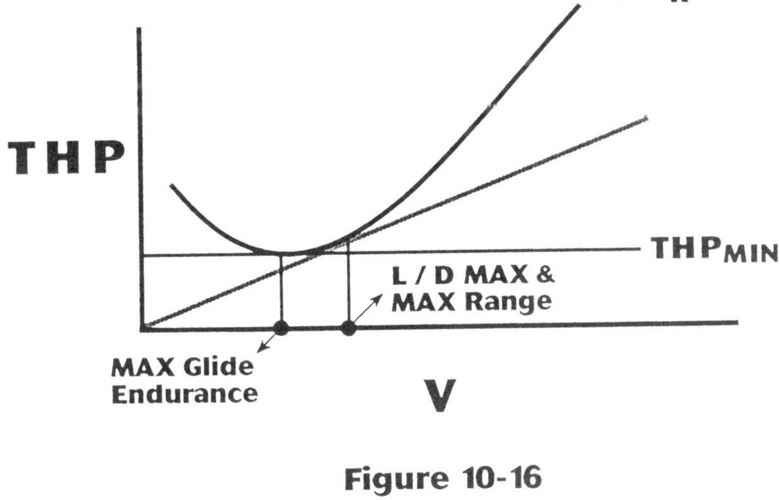

Figure 10-16

Prop / Jet

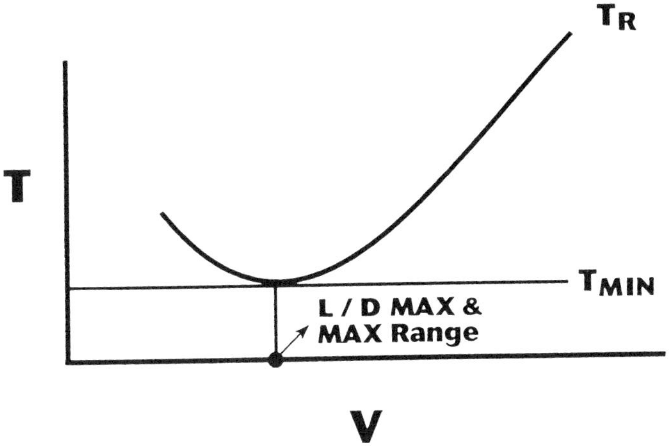

Figure 10-17

Study Questions Chapter 10

1. When will thrust equal drag?

2. When will the thrust required curve equal the total drag curve?

3. Define thrust required.

4. What happens if thrust does not equal thrust required?

5. What is the bottom point on the total drag/ thrust required curve?

6. Define thrust available.

7. What happens to the thrust available for a prop aircraft when there is an increase in velocity?

8. How is thrust available for a jet depicted and why?

9. Define excess thrust.

10. What happens if there is a positive value of excess thrust?

11. What is the intersection of the thrust required and thrust available curves?

12. What airspeed does $T_A - T_R = 0$ relate to?

13. What does a deficit value of excess thrust mean?

14. What fuel flow does the thrust available curve represent for a jet?

15. When does the minimum fuel flow occur for a jet?

16. What does the fuel flow in a jet do with an increase in altitude?

17. Depict the jet/ prop THP required curve.

18. What happens to the thrust horsepower available of prop as velocity increases?

19. What happens to the thrust horsepower available for a jet with an increase in velocity?

20. How is L/ D Max located on the thrust required curve?

21. How is L/ D Max located on the THP curve?

22. Is the velocity for L/ D Max the same on the THP and thrust curve?

23. Define maximum angle of climb.

24. Where does maximum angle of climb occur?

25. Where is the maximum angle of climb for a jet? for a prop?

26. What is the most commonly used angle of attack during a climb?

27. Where does the maximum rate of climb occur?

28. What velocity is a jets maximum rate of climb? A props?

29. How does the velocity of maximum rate of climb compare with the maximum angle of climb?

30. What does the pilot want when flying at maximum endurance?

31. How is fuel burned at maximum endurance?

32. What is a jet's fuel flow related to?

33. Where is the minimum fuel flow for a jet?

34. What graph is used to find maximum endurance for a prop?

35. What velocity is maximum endurance for a prop?

36. Define maximum range?

37. What happens to the velocity of a jet with a small increase in fuel flow?

38. Is maximum range for a jet greater than or less than the velocity for L/ D Max?

39. How does the maximum range for a prop compare with L/ D Max?

40. How does the maximum glide endurance airspeed for a jet and prop compare with L/ D Max?

41. How does the maximum glide range airspeed of a jet or prop compare with L/ D Max?

42. Which performance is not effected by increasing the weight of an aircraft?

43. Explain the effects of an altitude increase on aircraft performance.

44. Explain the effects of a headwind on aircraft performance.

Chapter 11

High Speed Aerodynamics

Definitions:

✈ Mach Number: Mach number is the ratio of TAS to the speed of sound at the altitude you are flying at. Speed of sound varies only with temperature.

Examples:

Altitude	Temperature	Speed of Sound
sea level	59F	661
10,000	23F	638
36,000	-70F	573

✈ Critical Mach Number (M Crit):
M Crit is the Mach number where supersonic flow first develops somewhere on the aircraft.

✈ Subsonic: Flight at speeds up to M Crit.

✈ Transonic: Flight at speeds from M Crit up to 1.2 to 1.3 Mach.

✈ Supersonic: Flight at speeds above 1.2 to 1.3 Mach.

Subsonic Flight:
Subsonic flight is flight at speeds up to M Crit.

→ Normal Shock Wave Formation:
(Reference Fig. 11-1)

At M Crit a normal shock wave forms which is at right angles to the free airstream. As the air flows through the normal shock wave:

1. V decreases, Ps increases, Density increases, and Temperature increases.
2. It maintains the same direction.
3. The loss of kinetic energy makes it difficult for the boundary layer to remain attached to the airfoil. This is called shock induced separation.
4. The higher pressure, density, and temperature create an additional drag in the airstream which increases total drag. This drag is called wave drag. Supersonic aircraft must have powerful engines to overcome wave drag.

→ Methods to Increase M Crit:
By increasing M Crit, the subsonic range is extended while the problems caused in the transonic range are reduced.

1. Thin Wings: Air doesn't accelerate as fast over the upper surface of thin wings, therefore, there will be a higher TAS and Mach number before M Crit is encountered. The problem with thin wings is that they are not suited for flight at slow airspeeds as is seen in their extremely high approach speeds.

2. Sweepback Wings: Sweepback reduces the amount of the relative wind which flows across the wing and is accelerated. This allows for a higher TAS and higher Mach number before M Crit is encountered.

Sweepback must exceed 35 degrees before a significant increase in M Crit takes place. Sweepback can increase M Crit .1 to .15 Mach. The problem with sweepback is the spanwise flow which causes the wing tip to stall. Stall fences can be used to reduce this.

Transonic Flight:
Transonic flight is flight at speeds from M Crit up to 1.2 to 1.3 Mach.

✈ Upper and Lower Shock Wave: The normal shock wave also forms on the bottom surface of the wing. The bottom wave acts the same as the upper but it is weaker.
(Reference Fig. 11-2)

✈ Shock Stall: Shock stall is caused by turbulence from shock induced separation which hits the tail and causes buffet similar to subsonic stall buffet.

✈ Normal Shock Waves at Trailing Edge: As the normal shock waves move aft, the effects of shock stall reduce because there is less shock induced separation.
(Reference Fig. 11-3)

✈ Force Divergence Mach Number: Force divergence Mach number occurs at velocities 5% to 10% above M Crit where the C_D takes a sudden upswing. The drag increases greatly at force divergence due to:
1.The energy losses across the normal shock wave.
2. Boundary layer separation.
(Reference Fig. 11-4)

✈ Oblique Shock Waves at Mach 1: When the aircraft speed reaches Mach 1, the upper and lower normal shock waves attach to the trailing edge and become oblique. Velocity reduces through an oblique shock wave but it remains supersonic.
(Reference Fig. 11-5)

✈ Bow Wave: At Mach 1 a bow wave forms on the leading edges. A high pressure area forms directly in front of a leading edge and is called the area of stagnation. Flow in the area of stagnation is subsonic. As long as the area of stagnation exists, the aircraft will remain in transonic flight. You can't get into supersonic flight until the bow wave attaches to the leading edges. If you have rounded leading edges (like the A6 Intruder), you will never get supersonic because there will always be the area of stagnation and the bow wave will never attach to the leading edges. If a leading edge is sharp the bow wave will attach and become oblique at the upper transonic range. A bow wave greatly increases wave drag.
(Reference Fig. 11-6)

✈ Aircraft Design Features to Reduce Transonic Flight Problems:
1. Thin Wings: Thin wings and sweepback reduce wave drag and increase M Crit.
2. Vortex Generators: Vortex generators reduce the amount and intensity of shock induced separation.
3. Boundary Layer Control: BLC devices blow air into the boundary layer to keep it at a higher energy level. This reduces shock induced separation and wave drag.

4. Slab Type Controls: Slab type controls decrease the adverse effects of trying to operate control surfaces in the turbulent air caused by shock waves.
5. Area Rule: Area rule is used to reduce wave drag. In the area rule the fuselage has a waist at the wing root junction. This allows the supersonic air to remain close to the fuselage and thus reduces wave drag. If supersonic aircraft did not have a waist, the supersonic air would compress and increase wave drag.
(Reference Fig. 11-7)

Supersonic Flight:
You have reached supersonic flight when the bow waves attach to the leading edges and all shock waves at leading and trailing edges become oblique.

✈ Characteristics of Oblique Shock Waves: With flow passage through an oblique shock wave:
1. Pressure, density, and temperature increase.
2. Velocity reduces but remains supersonic.
3. Flow direction changes so that the airflow remains parallel to the surface of the airfoil and therefore flow separation does not exist.

Changes through an oblique wave are not as drastic as through a normal shock wave, therefore drag is much less.
(Reference Fig. 11-8)

✈ Expansion Wave: An expansion wave forms where supersonic flow encounters a point where the cross sectional area increases (around a corner). As the

supersonic air expands through the expansion wave:
1. Pressure, density, and temperature decrease.
2. Velocity increases.
3. Direction changes causing the airflow to remain parallel to the airfoil and thus remain smooth.
(Reference Fig. 11-9)

✈ Shapes of Supersonic Airfoils:
(Reference Fig. 11-10)
This type of airfoil which causes smooth supersonic flow because of the expansion wave, cannot be used at subsonic speeds because subsonic flow cannot flow around the corners and thus flow separation would occur. The most realistic supersonic airfoil is one which is smooth, cambered, and thin. This type can fly at subsonic and supersonic speeds.

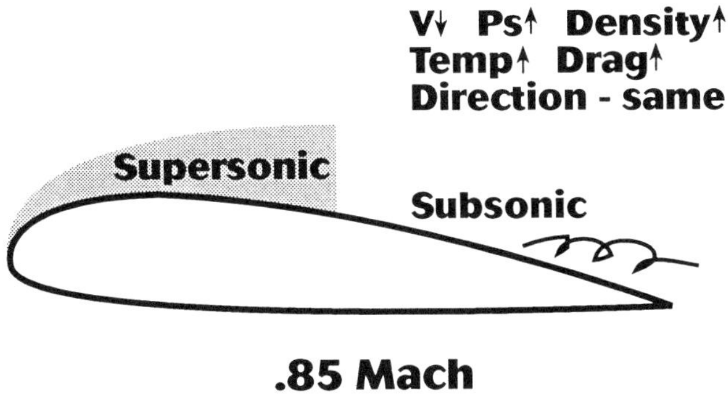

Figure 11-1

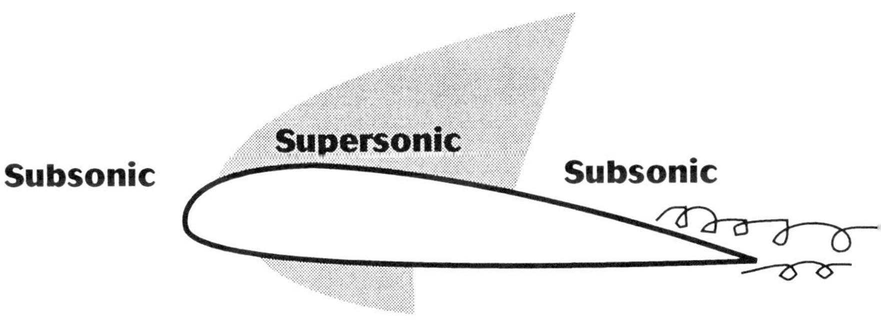

Figure 11-2

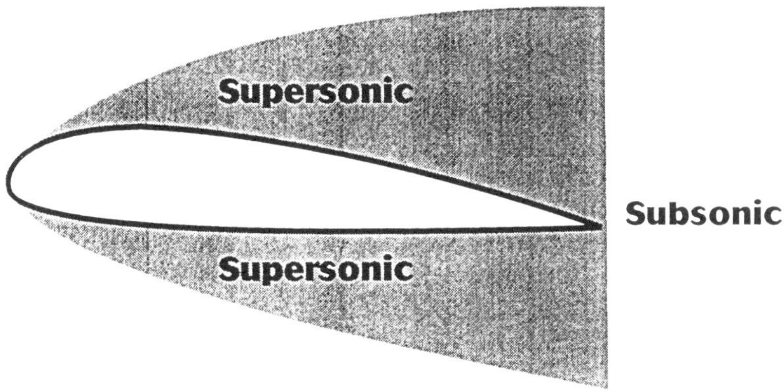

Figure 11-3

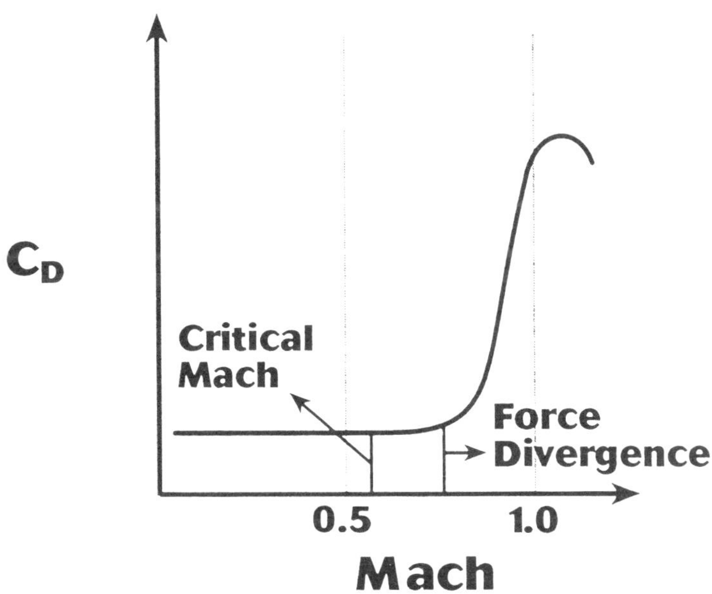

Figure 11-4

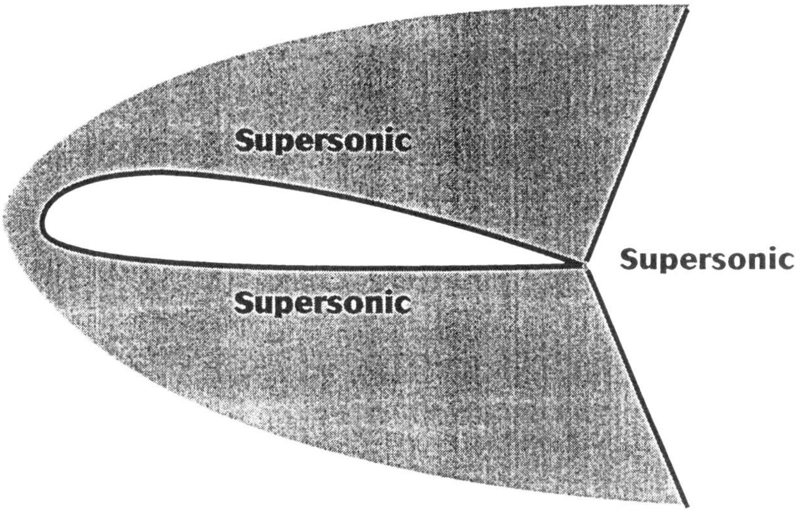

Mach 1

Figure 11-5

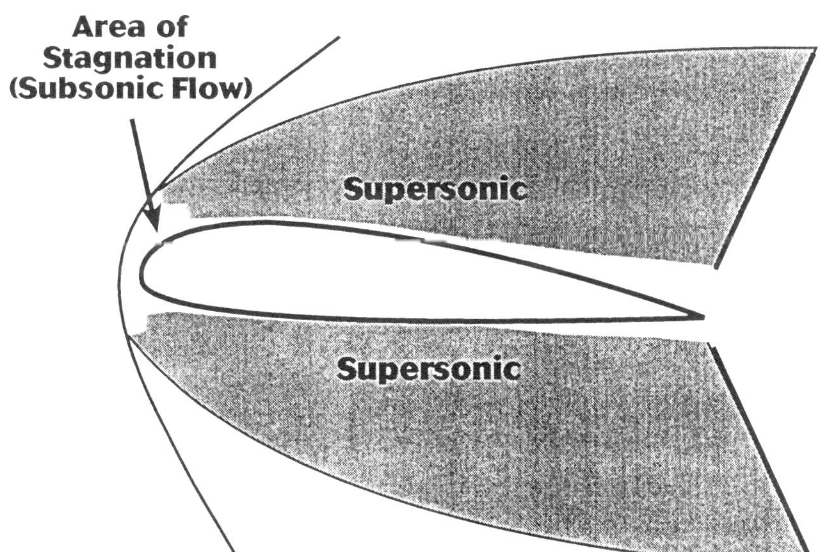

Mach 1

Figure 11-6

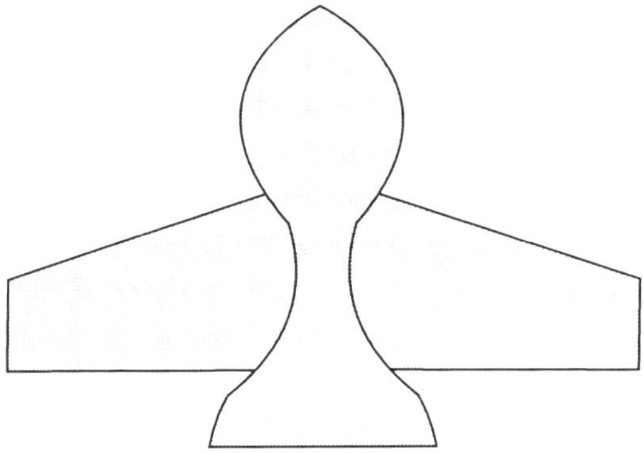

Figure 11-7

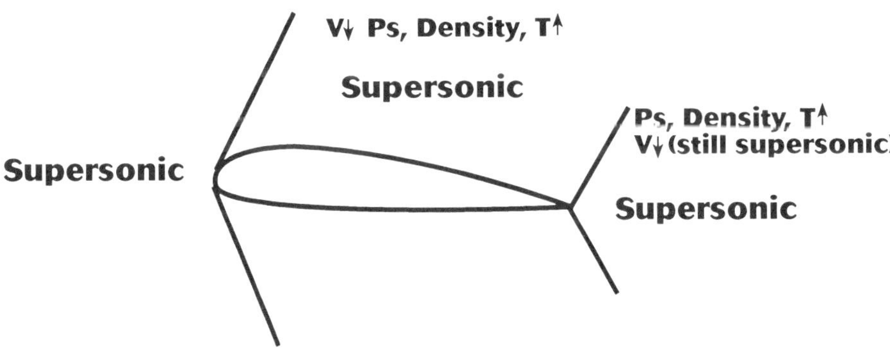

Figure 11-8

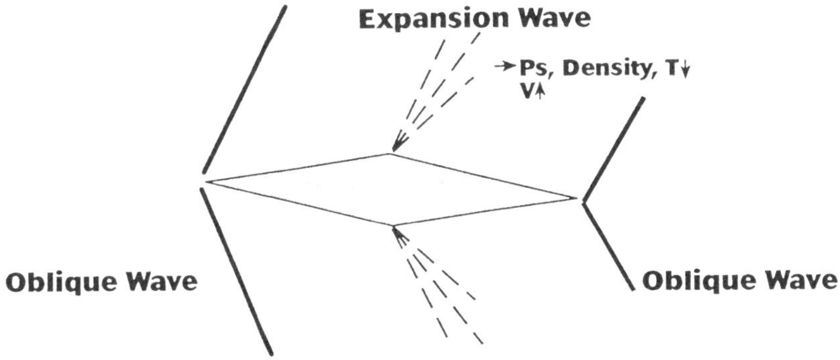

Figure 11-9

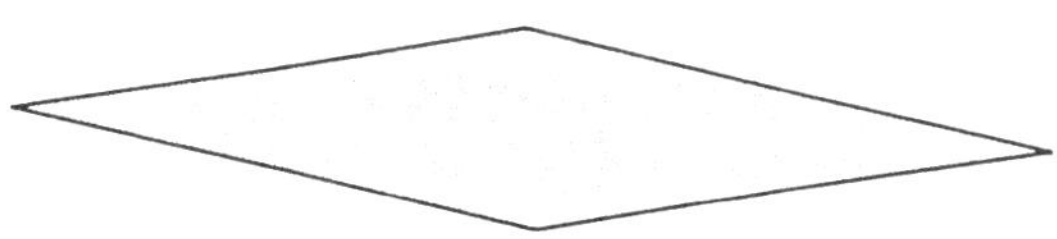

Figure 11-10

Study Questions Chapter 11

1. Define Mach number.

2. Define Critical Mach number.

3. Define subsonic flight.

4. Define transonic flight.

5. Define supersonic flight.

6. When does a normal shock wave form?

7. Why is it useful to increase M Crit?

8. Name two methods used to increase M Crit.

9. What is shock stall?

10. What happens to the effects of shock stall as normal shock waves move aft?

11. What happens to drag at the force divergence Mach number?

12. What shape do shock waves become when speed reaches Mach 1 and they attach to the trailing edge of an airfoil?

13. When does a bow wave form?

14. Can an aircraft become supersonic without the bow wave attaching to the leading edges?

15. Why are thin wings and sweepback used to reduce transonic flight problems?

16. Why do supersonic aircraft use slab type controls?

17. What is area rule used for?

18. When does an aircraft obtain supersonic flight?

19. What is the most realistic shape of a supersonic airfoil?

Index

A

B

C

D

E

F

G

H

I

L

M

N

P

R

S

T

U

V

W

References and Suggested Reading

Hurt, H. H. (1965). Aerodynamics for Naval Aviators. Navair (00-80T-80).

Smith, Hubert "Skip" (1985). The Illustrated Guide to Aerodynamics. Tab Books, Inc.

Hubin, W. N. (1992). The Science of Flight. Ames, IA: Iowa State University Press.

Federal Aviation Administration. (1980). Flight Training Handbook. Basin, WY: AMP Publishers.

Federal Aviation Administration. Stall and Spin Awareness Training. Advisory Circular 61-67B. Washington, DC: Author.

Federal Aviation Administration. Aircraft Wake Turbulence. Advisory Circular 90-23E. Washington, DC: Author.

Federal Aviation Administration. (FAA P-8740-48). On Landings Part 1. Washington, DC: Author.

Federal Aviation Administration. (FAA P-8740-49). On Landings Part 2. Washington, DC: Author.

The Pilot's Outline Guide to Basic Aerodynamics

Jones, ISBN 0-9639984-2-0

Number of copies @ $12.95 each ______
Shipping and handling included for orders in the U.S. and Canada. Orders outside this area please add 10% of total order to help cover shipping.

Total Enclosed. ______________

Ship to:

NAME ______________________________

ADDRESS ___________________________

____

Telephone _________________________

Mail check or M.O. payable to:

Eastern Dakota Publishers
P.O. Box 12512
Grand Forks, ND 58208-2512

E-Mail orders or information:
EDPUBLISH@aol.com

* Write for volume order discounts.